DATE
DISCARD

Big Dish

University Press of Florida · State University System
UWF
FSU
FAMU
UNF
UF
UCF
USF
FGCU
FAU
FIU
Florida A&M University, Tallahassee
Florida Atlantic University, Boca Raton
Florida Gulf Coast University, Ft. Myers
Florida International University, Miami
Florida State University, Tallahassee
University of Central Florida, Orlando
University of Florida, Gainesville
University of North Florida, Jacksonville
University of South Florida, Tampa
University of West Florida, Pensacola

Douglas J. Mudgway

BIG

UNIVERSITY PRESS OF FLORIDA
GAINESVILLE
TALLAHASSEE
TAMPA
BOCA RATON
PENSACOLA
ORLANDO
MIAMI
JACKSONVILLE
FT. MYERS

DISH

Building America's Deep Space Connection to the Planets

Printed in the United States of America on recycled, acid-free paper

10 09 08 07 06 05 6 5 4 3 2 1

Library of Congress Cataloging-in-Publication Data
Mudgway, Douglas J., 1923–
Big dish : building America's deep space connection to the planets /
Douglas J. Mudgway.
p. cm.
Includes bibliographical references and index.
ISBN 0-8130-2805-1 (alk. paper)
1. Deep Space Network. 2. Outer space—Exploration—United States.
I. Title.
TL3026.M82 2005
621.382'54—dc22 2004061154

The University Press of Florida is the scholarly publishing agency
for the State University System of Florida, comprising Florida A&M
University, Florida Atlantic University, Florida Gulf Coast University,
Florida International University, Florida State University, University
of Central Florida, University of Florida, University of North Florida,
University of South Florida, and University of West Florida.

University Press of Florida
15 Northwest 15th Street
Gainesville, FL 32611-2079
http://www.upf.com

To Bill Merrick and the Hard Core Team,
who realized the vision of others,
and to those who followed,
who sustained and nourished it.

Contents

9. Implementation 176

10. Closing Scenes 200

Foreword

To some, the U.S. space program has represented prestige and a positive image for the United States on the world stage. To others it has signified the quest for national security. Some others view it solely as the realm of telecommunications satellites and little else. To still others space is, or should be, about gaining greater knowledge of the universe; it represents, for them, pure science and the exploration of the unknown. Even so, the history of space science and technology is one of the largely neglected aspects in the history of the space program. This important book by Douglas Mudgway describes an essential element of that technology, NASA's great communications links to deep space. Without this deep space connection, the wonderful missions and space probes that have been sent out from Earth to explore the other bodies of the solar system would not have been possible.

Since 1958, when NASA was first established, every planet of the solar system save Pluto has been explored in a modest way at least once, and several of the moons of planetary bodies (including our own) and some comets and asteroids have been visited. We have placed spacecraft in orbit around our moon and the planets Venus, Mars, and Jupiter, and have landed on Venus, Mars, and the Moon. NASA's stunning missions to explore the outer solar system have yielded a treasure of knowledge about our universe, how it originated, and how it works. NASA's exploration of Mars, coupled with the efforts of the Soviet Union/Russia, has shown powerfully the possibility of past life on the Red Planet. Missions to Venus—including some that landed on it—and to Mercury have harvested understanding about the inner planets. Lunar exploration has exponentially advanced human knowledge about the origins and evolution of the solar system. Most important, we have learned that Earth is a place in which everything necessary to sustain life is "just right," while all the other planets of our system seem exceptionally hostile.

Planetary exploration has not taken place by magic. It required visionary leadership, strong-willed management, and persevering execution. Like NASA and other aspects of the space program, it began as a race between the United States and the Soviet Union to see who would be the first to get some sort of spacecraft near the Moon. It expanded in the 1960s, when space science first became a major field of study. During that decade both the United States and the Soviet Union began an impressive effort to gather information on the planets of the solar system using unmanned robotic spacecraft.

The studies based on this new data have revolutionized humanity's understanding of Earth's immediate planetary neighbors. These studies of the planets captured the imagination of people from all backgrounds and perspectives. Photographs of the planets and theories about the origins of the solar system appealed to a very broad cross section of the public. As a result, NASA had little difficulty in capturing and holding a widespread interest in this aspect of the space science program.

The dazzling products of the first forty-five years of NASA's lunar and planetary exploration program were made possible by brilliant, innovative, and dedicated people who conceived, built, and operated exquisite machines that turned their visions into reality.

Some of these machines were called spacecraft. Boosted by powerful rockets into orbits beyond the influence of Earth's gravity, they could and did carry their payloads of scientific instruments to the farthest reaches of the solar system and faithfully transmitted back to Earth the scientific marvels that they found there.

Other machines remained on Earth to maintain the deep space connection by means of which controllers, engineers, and scientists gave sense and direction to their distant robotic craft, and by which the robots' harvest of scientific bounty was returned to Earth. These latter machines, themselves consummate examples of advanced technology, were the giant antennas that formed the basis of NASA's "deep space connection."

This book is their story.

Roger D. Launius
National Air and Space Museum
Smithsonian Institution, Washington, D.C.
September 25, 2003

Acknowledgments

I was assisted in the writing and publication of this book by a large number of others for whose collective help I am profoundly grateful.

Among them were some who were directly involved, from the very beginning of the Advanced Antenna Project, in the events recorded here and who gave unreservedly of their time, technical expertise, and enthusiasm to my purpose of writing the narrative. They accorded me personal interviews and provided me with private documents, draft reviews, and answers to endless technical questions. For these specific reasons I am deeply indebted to Dan Bathker, Ronald Casperson, Agustín Chamarro, Robert Clauss, José Fernandez, Robert Hall, Donald McClure, Houston McGinness, Fred McLaughlin, Horace Phillips, Charles Stelzried, Robertson Stevens, José Urech, Chris Valencia, and Dale Wells.

Regrettably, we missed Bill Merrick, whose wonderful personality we well remembered. He passed away in 1997 at 80 years of age, but his daughter, Beth Merrick, graciously provided me with some background material on his non-professional life.

In Australia, the director of the Canberra Deep Space Communications Complex, Peter Churchill, provided a current view of the status of the 70-meter antenna at that location, while Sarah Bugg provided historical details of the Tidbinbilla site. Additional details of the ancient aboriginal association with Canberra came from Lois Padgam of the Tidbinbilla Nature Reserve.

In Spain, Agustín Chamarro, former director of the Madrid Deep Space Communications Complex, assembled a detailed account of the association of the Robledo de Chavela pueblo and surrounding area with Spanish history dating back to the sixteenth century.

At Goldstone, Don Mischel, former director of the Goldstone Deep Space Communications Complex arranged many visits to that site which enabled me to become thoroughly acquainted with the inner workings of the 70-meter antenna.

Invaluable assistance with archival research at the Jet Propulsion Laboratory was provided by Shirley Wolff and by John Bluth and Charles Miller at the JPL Archival and Records Facility. Fred McLaughlin's beautiful illustrations clarified the subject matter and enhanced the narrative at salient points throughout the book. Jane Odom and Steve Garber of the NASA History Office assisted me with archival research at NASA headquarters. Maybelle Lipking and Germaine Moon of the Mojave River Valley Museum at Barstow led me to fascinating details of the historic past of the Barstow area, while Neil Morrison of the U.S. Army's National Training Center Museum at Fort Irwin shed light on the military association with the Goldstone Dry Lake area.

My original manuscript benefited immensely from the insightful comments and helpful suggestions of David Whalen and others, who generously undertook its review. I am most indebted to them for their help.

Under the direction of John Byram, the editorial and production staff at the University Press of Florida gave the book form and substance, and to them also an expression of gratitude is due.

Finally, I need to formally recognize NASA/JPL and NASA/JPL/University of Arizona for the use of their magnificent images, and credit NRAO/AUI for the Tatel antenna and the impressive picture of the Green Bank Telescope.

Introduction

The Challenge

In mid-1958, as the United States quickened its response to the Soviets' *Sputnik* challenge for technological supremacy in space, a group of engineers toiled frantically to erect a huge dish-shaped radio antenna in a remote area of California's Mojave High Desert. This was no ordinary antenna. Its aluminum surface, 85 feet (26 meters) in diameter, was parabolic and would act like an optical telescope to focus radio waves onto sensitive receivers carried on a tripodlike structure above its rim. Its extremely rigid supporting structure was designed to precisely track the faint radio signal from a yet-to-be-launched space probe as it moved from horizon to horizon. Later, complex electronic equipment located nearby would extract the data carried by the radio signal and convert it into human-readable form, for analysis and computation by scientists and engineers at JPL—the Jet Propulsion Laboratory of the California Institute of Technology in Pasadena.

Earlier that year, engineers from JPL had captured the nation's attention with a riveting demonstration of U.S. technology that countered *Sputnik*'s launch by setting a rather more sophisticated satellite named *Explorer* into orbit around the Earth. That exploit evened the score with the Soviets, but JPL wanted to do more. Its New Zealand–born director William Pickering now sought to demonstrate America's prowess in space-related technology even more convincingly. He was aiming for the Moon, and he had two small spacecraft named *Pioneer 3* and *Pioneer 4* to do it with.

What he did not have was a communications link of sufficient sensitivity to track the faint radio signal from a Pioneer spacecraft as it sped ever further away from Earth toward the Moon. This radio "voice" carried important data about the well-being of the spacecraft and its speed and acceleration. It also carried the science data from James Van Allen's Geiger-Mueller counters that would measure the density of belts of radiation that surround the Earth. It was imperative that the radio tracking station be ready and working properly before the *Pioneer 3* launch, then firmly scheduled for December.

In the event, Bill Merrick and his team completed the antenna just in time, and *Pioneer 3* was launched from Cape Canaveral, Florida, to the Moon on December 6, 1958. It was the first mission to be launched under the aegis of the partnership between JPL and the National Aeronautics and Space Administration (NASA), a partnership that would endure with great success far into the future. Although *Pioneer 3* did not reach the Moon, the data it transmitted back to the tracking station in the Mojave contributed to the major scientific discovery of dual bands of radiation, called the Van Allen Belts, that encircle the Earth.

Pioneer 4 was launched in March of the following year, and passed by the Moon about 29 hours later. The antenna in the Mojave tracked its faint radio signal for 82 hours, out to a distance of 407,000 miles (655,000 kilometers), before losing contact. No human-made object had ever been tracked to that distance before. The probe transmitted excellent data on the radiation environment in space and eventually settled into a heliocentric orbit, becoming the first U.S. spacecraft to orbit the Sun (Siddiqi 2002).

The U.S. space program, now directed by NASA, was gaining momentum, and Pickering of JPL was becoming a key player. JPL had by then been elected to lead NASA's Explore the Solar System initiative, and Pickering was determined to do his exploring where none had gone before, beginning with close-up pictures of the lunar surface. The program was called Ranger.

Targeted for the Moon, the Ranger spacecraft were equipped with television cameras and radios intended to transmit a series of close-up images of the lunar surface back to Earth in the last few minutes before impact. Considerably more complex than the Pioneer probes, the Ranger craft would also require constant monitoring of their "vital signs" throughout their Earth–Moon journey. This meant that more than a single tracking station would be needed to maintain contact with a spacecraft as the Earth rotated beneath it.

Using its international goodwill, the United States was able to negotiate treaties with Australia and South Africa that permitted NASA to construct space-related facilities in those countries. In a short space of time, NASA/JPL built two more antennas of similar size and design to the antenna in California, one at Woomera in South Australia and the other near Johannesburg in South Africa, to form a ring of tracking stations at 120-degree intervals around the Earth. Telephone and teletype lines linked each station to a control center at JPL in Pasadena. With such a network of antennas, controllers at JPL could maintain round-the-clock contact with spacecraft departing Earth on journeys of any duration. Telemetry and science data could flow in an unbroken stream from spacecraft traveling anywhere in deep space to controllers at JPL via the far-flung antennas of the DSN. So was born America's Deep Space Network.[1]

Although it started off badly, the Ranger program was eventually crowned with success and delivered many high-resolution pictures of the lunar approach that were up to a thousand times better than could be obtained from Earth-based cameras of the time. These images were used to determine future Apollo landing sites.

Even as the two new overseas antennas were being completed to support the first of the Ranger missions in 1961, engineers at JPL were looking beyond the Moon, much further out into deep space, toward the planets. JPL was proposing a new family of spacecraft to explore the planets, first the morning star Venus and later the red planet Mars. They would be called the Mariners and would be designed to take pictures and make scientific observations of a planet's surface as they flew close by.

However, ambitious new missions like these brought with them a host of new technical problems, among them how to extend the existing DSN communications links from lunar distance out to Venus and to Mars, distances that were 500 to 1,000 times greater. DSN engineers found innovative ways to enhance the performance of the existing 26-meter antennas, and the Mariner missions went forward. By 1964, telemetry data transmitted by the Mariners to JPL's three 26-meter antennas had given the world its first close-up views and new insight about the nature of Earth's nearest neighbors.

By that time, a kind of synergistic relationship had developed at NASA/JPL between the planetary spacecraft community and the deep space communications community. No sooner did the communications engineers find a way to enhance the capability of the ground antennas and receivers than spacecraft engineers conceived a new mission to take advantage of the new communications capability, and vice versa. They were heady days for the esoteric world of deep space communications.

Out of that environment emerged a remarkable new DSN initiative that, at its conception in 1962, promised to give the United States space program an unassailable lead in the "race for space." Down the corridors of power at NASA headquarters it was called the Advanced Antenna System or AAS, but at JPL it came to be known as the Big Dish project.

Although the original antennas of the DSN were considered large for structures of that type, the Big Dish would be gigantic. With a diameter of 210 feet (64 meters), a surface area of almost one acre, and a weight of 8,000 tons, the proposed antenna would dwarf a modern nine-story office building. Yet the shape of its precision reflector surface would be within .25 inch (6.35 mm) of a perfect parabola. It would float on a film of oil the thickness of a few sheets of computer paper, and could be pointed to any direction in space with unprecedented accuracy. With such an instrument the DSN would be able to communicate with spacecraft traveling to Venus and Mars, to Jupiter, even to Saturn if JPL could build a spacecraft that could reach that deeply into space.

The JPL spacecraft engineers responded to the challenge. They built not one but two spacecraft that could, and did, reach Saturn. They were called the Voyagers. And the DSN built not one but three 64-meter antennas. Completed in 1968, they were located at remote sites in California, southeastern Australia, and central Spain. Working together, the Voyagers and the DSN's 64-meter antennas brought the world an entirely new vision and body of scientific knowledge and understanding of the solar system and its fabulous constituents. Absolutely dazzling images—Mars's landscapes, Jupiter's Giant Red Spot and horrendous atmospheric vortices, Saturn's ephemeral rings and intriguing satellites— flowed in a seemingly never-ending stream from the Voyagers through the DSN antennas to the exhilarated scientists at JPL and to the public at large. NASA/JPL was on a roll, and there was no longer a contest for the accolades of deep space.

But there was more to come. After passing Saturn, the Voyagers kept going, and Uranus and Neptune beckoned. "More, we want more science," cried the scientists. Once again the communications links would have to be stretched, this time to unimaginable distances. Uranus was twice as far from Earth as was Saturn, and Neptune was three times as far, thirty times the distance from Earth to the Sun. That was really "deep" space. To the innovative engineers and scientists at JPL it was an irresistible challenge, and they met it. In a most remarkable feat of precision large-scale engineering, the DSN found a way not only to enlarge the original Big Dishes but also to enhance even further their capabilities. When the work was completed in 1988, the DSN's 70-meter-diameter antennas were not only the largest but also the best of their kind in the world. They were unique, and soon became indispensable to JPL's mission to explore the solar system.

Big Dish Dividend

The Planets

The scientific rewards from NASA's planetary exploration program were truly great. "A surfeit of science from a plethora of planets," one writer called them. For the Deep Space Network it was a "Big Dish dividend."

The technology development that produced the Big Dish dividend was driven by five major planetary programs over the years 1961–2001 (Mudgway 2001). The first 64-meter antenna came into service at Goldstone in the Mojave Desert about midway through the Mariner era (1961–74). It was immediately involved in the missions of *Mariner 4* to Mars and of *Pioneer 6* and *Pioneer 7* to explore fields and particles in the interplanetary medium. New missions were quickly added to its schedule—Surveyor moon landers, Lunar Orbiter moon orbiters, two additional Mariners to Mars and one to Venus, and several additional Pioneer missions including one to make the first flyby of Jupiter. By the time the Mariner era closed in 1974, two more 64-meter antennas had been added to the Deep Space Network, at Canberra and Madrid, the Apollo program was looking to the DSN 64-meter antennas for backup support to its own network of 26-meter antennas, and NASA was preparing to launch two massive spacecraft for the first attempt at a landing on Mars. The Viking era was at hand, and the 64-meter antennas were a keystone of the mission design.

The Viking era (1974–78) saw the first Mars landings accomplished successfully, in 1976, and the introduction of the first non-NASA mission into the DSN 64-meter tracking schedules. The spacecraft was named *Helios,* it was German in origin, and its mission was to explore the solar environment from a deep space orbit around the Sun. The Pioneer missions continued to return useful new science data from deep space, and two new Pioneer missions appeared, one to map the surface, the other to probe the atmosphere of Venus. In 1977, toward the end of the Viking era, the two Voyager spacecraft began a long and unpredictable journey into deep space, the first leg of which was Earth-to-Jupiter. It was the Voyager mission that would later define the nature of things to come in the DSN, particularly the way in which other antennas were used to supplement the awesome capabilities of the 64-meter antenna itself.

The magnificent Voyager mission dominated the DSN scene from about 1977 through 1986. That period was appropriately identified as the Voyager era. First Jupiter was visited (1979), then Saturn (1980–81), and then Uranus (1986). There seemed to be no limit to what could be done to extend the gathering power of the 64-meter antennas as the strength of the Voyager signals diminished with ever-increasing distance between spacecraft and Earth. Employing its own 34-meter antennas and with the cooperation of CSIRO (Commonwealth Scientific and Industrial Organisation) and NRAO (National Radio Astronomy Observatory) to make their large antennas available at specific times, the DSN was able to increase its receiving sensitivity to compensate for the deficit due to increased distance of the spacecraft from Earth. And Voyager, astounding as it was, was not the only mission that demanded attention from the 64-meter antennas. A lot of effort was expended on planning 64-meter support for Venus-Balloon/Pathfinder, a high-profile Russian attempt to land small spacecraft on Venus that eventually proved unsuccessful.

In anticipation of future demands for improved data-handling capability, the DSN underwent major changes in the Voyager era. The original 26-meter antennas were upgraded to 34-meter diameter, and improvements were made to the pedestal and alidade of the 64-meter antennas where necessary. Data processing facilities at the three remote sites were replaced with newer high-performance hardware and software that reflected advanced technology in all areas. The principal driving force for development and expansion of the network changed gradually, and sometimes with considerable overlap, from one major mission to the next.

About 1984, NASA's approach to launch systems for planetary spacecraft changed from expendable launch vehicles to space-shuttle-based launch vehicles. The first planetary spacecraft launched from a space shuttle was to be *Galileo,* for a mission to Jupiter in 1986.

Early studies had shown that, when the Galileo spacecraft reached Jupiter, the mission would require considerably more communications link capability than could be provided by the 64-meter antennas alone. To meet the Galileo need, the DSN planned to enlarge the diameter of the 64-meter antennas to 70 meters in 1987–88 and to array them with antennas of other agencies—one at Parkes, Australia, and a VLA (Very Large Array) at Socorro, New Mexico—when necessary. In 1986, however, before the work could be completed, *Challenger* exploded and all planetary launches were placed on hold. In the aftermath of the *Challenger* accident, *Galileo* was deferred to 1989. Then, like the *Magellan* radar-mapping mission to Venus, *Galileo* was launched into a planetary orbit by a solid-propellant launch vehicle carried into Earth orbit by the space shuttle. The solar polar explorer *Ulysses* followed in 1990. Meanwhile *Voyager 2* had reached Neptune, and the new 70-meter antennas, arrayed with Parkes and the VLA, were required to return the imaging and science data at such extreme range. By then the DSN's Big Dishes had been a crucial part of scientific missions to all of Earth's companions in the solar system except Pluto. All future missions would build upon the scientific knowledge gained from earlier visits, to further explore planetary atmospheres, surface composition, or global topography and climate.

They would use probes, landers, orbiters, and mappers, and all would depend upon the Big Dishes to achieve their scientific objectives.

In 1988 the Goldstone antenna was engaged in an international cooperative mission to Mars called Phobos. Despite high expectations for a successful mission, both spacecraft were lost near Mars the following year. A similar fate awaited *Mars Observer* in 1993. Although all of these missions used the DSN's smaller 34-meter antennas for their near-Earth phases, they all depended upon the 70-meter antennas to capture the science data that was to be transmitted from their ultimate destinations. *Galileo* suffered a setback in 1991 when its high-gain antenna failed to deploy, and the DSN was called upon to participate in a joint effort with spacecraft engineers to recover the mission. That effort was successful, and the mission went on to achieve at least 70 percent of its original scientific objectives before the primary mission ended in 1996. Under the new mission name Europa, *Galileo* continued to return data from Jupiter with particular emphasis on investigating the composition and origin of Jupiter's satellite Europa for possible evidence of life forms.

NASA/JPL spacecraft made two spectacularly successful return visits to Mars in 1996. *Mars Global Surveyor* and *Mars Pathfinder/Rover* arrived at Mars in November and December respectively, and immediately set about their tasks of exploring the Martian surface. The former mapped the entire Martian surface from orbit, while the latter landed on the surface and carried out close-up imaging and spectrographic observations of surface features as well as meteorology observations of the Martian atmosphere.

The dominant mission of this era was, however, the huge Cassini mission to Saturn. Cassini was a follow-up to Pioneer's brief encounter with Saturn in 1979 and the Voyagers' more detailed observations in 1980–81. It was seen as the next logical step in the exploration of the solar system after the Galileo mission to Jupiter, and was a cooperative effort by NASA, the European Space Agency, and the Italian Space Agency. Equipped with twelve scientific experiment packages, the spacecraft *Cassini* was to orbit Saturn for a four-year period to study the Saturnian system in great detail. Like *Galileo, Cassini* carried a small atmospheric probe that, as it descended by parachute after release from the orbiter, was to sample and determine the composition and structure of the thick atmosphere of Titan.

By 1996 NASA had reverted to expendable launch vehicles for planetary missions, and consequently *Cassini* began its long journey of 3.5 billion kilometers (2.2 billion miles) from Cape Canaveral to Saturn aboard a Titan/Centaur launch vehicle. Like *Galileo, Cassini* was designed to use several gravity assists to reach Saturn. The sequence of gravity assists read like the stops on a bus route:

Figure 1.1. *Cassini* at Jupiter. *Cassini* captured this beautiful image of Io against the swirling cloud tops of the Jupiter atmosphere shortly after its closest approach to the planet in December 2000. The foreshortening effect can be misleading. Although Io appears to be close to the cloud tops in this picture, it is in fact about two and one-half Jupiter diameters above them. After leaving the environs of Jupiter early in 2002, *Cassini* continued its six-and-one-half-year journey to Saturn, where it arrived on July 1, 2004. NASA/JPL.

Venus 1, April 1998; Venus 2, June 1999; Earth, August 1999; Jupiter, December 2000; Saturn Arrival, July 2004; Titan Entry, November 2004. The average communication distance from Saturn to Earth would be 1.43 billion kilometers (890 million miles), and of course *Cassini* depended on the 70-meter antennas to bridge the communications gap.

Following a successful launch and completion of the Venus and Earth gravity assist maneuvers, *Cassini* reached Jupiter in December 2000 as planned. The beautiful picture shown in figure 1.1 leaves a permanent record of *Cassini*'s passage through the Jovian system.

The complete spacecraft stood 6.7 meters (22 feet) high, was 4.0 meters (13.1 feet) wide, and weighed 5,712 kilograms (12,593 pounds). The mission was to cost a total of U.S. $3.3 billion. But *Cassini* was the last of the monster spacecraft.

By the time *Cassini* was launched, winds of change had swept though NASA and "Faster, better, cheaper" became a familiar slogan. The Discovery and New Millennium programs were initiated and quickly produced small, high-efficiency spacecraft that were designed to carry out planetary missions with limited objectives at much less cost. *Mars Pathfinder/Rover* and *DeepSpace 1* were examples. The 70-meter antennas continued to be the anchor to which all of these missions were tied, of course, but the DSN was also challenged to match the economies of scale in the field of spacecraft design with corresponding economies in the operating costs of the network.

A new era was at hand.

Planetary Radar

In mid-March 1961, Eberhardt Rechtin sent a telex message to colleagues at MIT's Lincoln Laboratory: "Have been obtaining real-time radar signals reflected from Venus since March 10 using 10 kw CW [Continuous Wave] at [a frequency of] 2388 MC at a system temperature of 55 degrees [Kelvin]" (Butrica 1996). Rechtin was reporting the successful outcome of experiments that he, Walter Victor, and Robertson Stevens were conducting at Goldstone with the first 26-meter antennas that would later become part of the Deep Space Network. For some time previously a team at Lincoln Laboratory had been working on a similar experiment, but so far their effort had not succeeded. The data from the JPL experiments eventually led to an important scientific advance—namely, refinement of the then-current value for the astronomical unit (AU). Thus was born the concept of using the Deep Space Network as a scientific instrument in its own right. That idea was subsequently nurtured by the pioneering work in planetary radar of Richard M. Goldstein, a young JPL scientist who was also part of the Venus experiment team.

While the primary function of the DSN was to capture the science data generated and transmitted by a planetary spacecraft, the DSN was also interested in collecting other types of science data for researchers in other fields. One such field was radio science, another radio astronomy. Radio science depended upon a complex analysis of the way in which the radio-frequency downlink from a planetary spacecraft was disturbed as it traversed a planetary or solar atmosphere between spacecraft and Earth. Radio astronomy, on the other hand, did not require a spacecraft—it depended upon the radio emissions from extragalactic radio sources such as radio stars and quasars. For studying passive targets like planets, comets, and asteroids, it was necessary to illuminate the target with radiation transmitted from Earth, in which case the science was called radar astronomy.

Although the 26-meter antennas had been used effectively for research in all three areas since the first Venus experiment in 1961, investigators were quick to avail themselves of the tenfold improvement in observing capability that resulted from the introduction of the 64-meter antennas in 1966. Suddenly interest in all three fields of research expanded rapidly.

First on the scene was Richard Goldstein. Late in 1965, before the Goldstone antenna was yet fully operational, Goldstein used the antenna to observe the spectral spreading of the downlink signal from *Mariner 4* as it passed through the solar corona. This experiment was the forerunner to a large body of research that was subsequently carried out with the Pioneer, Viking, Helios, and Magellan spacecraft. Goldstein's main interest, however, was in the use of the Goldstone antenna for extending his ongoing planetary radar observations, and he quickly installed the necessary transmitting and receiving and data processing equipment for that purpose. Using the power of the 64-meter antenna and its enhanced radar capability, Goldstein and others turned their attention again on Venus and studied the surface in much greater detail than had hitherto been possible.

In 1968 they detected an Earth-crossing asteroid named Icarus. The report of this work was the first of many to be published in the distinguished scientific journal *Icarus.* Two years later, in a joint experiment with MIT's Haystack Antenna, the Goldstone radar detected reflections from Jupiter's satellites Ganymede and Callisto. The rings of Saturn were detected by radar on numerous occasions between 1974 and 1979 to complement the Voyager encounters with Saturn. In 1975 and 1976 the radar group at JPL made observations of Mars to determine surface roughness and slopes of potential landing sites for the Viking landers. Following the successful radar detection of Icarus in 1968, interest in radar detection of asteroids increased. Over the next several years, techniques to characterize their size, shape, and surface properties were improved, and additional asteroids were detected and characterized.

In 1983 the DSN established a Science Office to coordinate the science community's overwhelming demand for observing time on the 64-meter antennas for radio astronomy and radar observations, and for a program of investigation related to Earth crust motions known as Crustal Dynamics. Allocation of observing time on the 64-meter antennas for ground-based science experiments was subjected to the same rigor as that applied to the in-flight projects. By then the so-called Goldstone Solar System Radar (GSSR) was sorely in need of an upgrade to its capability and a new direction for its future field of research. By 1988, when the Goldstone antenna was upgraded to 70-meter diameter for the Voyager encounter with Neptune, the GSSR got its upgrade and, under the direction of a new manager, Steven J. Ostro, began a productive program of detection, characterization, and three-dimensional modeling of asteroids. Scientific interest in radar observations of planets and their rings and satellites had by then diminished, and for the Goldstone 70-meter antenna, the future trend of planetary radar pointed toward this new field of investigation.

Radio Astronomy

In the course of a 1997 interview at JPL, Michael Klein, then manager of the DSN Science Office, opined, "The DSN is recognized as a world-class instrument for radio astronomy, and allows astronomers to take advantage of the very latest technology in Earth- and space-based interferometry to produce first-rate science. The close symbiotic relationship that has developed between technology in the DSN and science in the DSN serves as a valuable stimulant to the benefit of both. In addition," he pointed out, "a strong constituency for science in the DSN promises improvements in both technical and cost performance for the DSN, and enhances the image of NASA in the international scene by virtue of the significance of its contributions to the sciences of radio and radar astronomy" (Mudgway 2001).

The body of work to which he referred was the outcome of a generous NASA policy that allowed the science community to make use of NASA facilities, specifically the 64-meter/70-meter antennas, for carrying out qualified scientific experiments during periods of time when the antenna was not otherwise being used for its primary task of tracking spacecraft. Interest in radio astronomy expanded rapidly in 1966 when the Goldstone 64-meter antenna first became available. By 1969 the demand for observing time on the new antenna far exceeded the time available for radio astronomy purposes. To make best use of the limited resources, a panel of experts assigned priorities based on scientific merit to all experimenters' proposals, and observing time was allocated accordingly. Under this system, radio astronomy figured importantly in DSN activity for the next twenty years.

During this time the 64-meter antennas in California, Spain, and Australia were used, either by themselves or in conjunction with adjacent 26-meter/34-meter antennas, to investigate extragalactic radio sources, Jupiter radio emissions, quasars, pulsars, galactic nuclei, interstellar microwave emissions, and many other subjects of great scientific interest to astronomers. In addition to those sponsored by NASA, proposals from other agencies such as the National Science Foundation, Caltech, and CSIRO were accepted and allocated observing time on the antennas. Cooperative experiments with radio astronomers in Australia, Spain, Italy, Japan, Sweden, Russia, Germany, and South Africa enhanced NASA's image in the international science community.

In the 1960s radio astronomers developed a powerful technique for creating a radio beam of extremely narrow angular width by suitably combining the signals from two (or more) widely separated antennas. Known as Very Long Baseline Interferometry or VLBI, it made its first appearance in the DSN in 1967 when the 64-meter antenna at Goldstone was combined with the 26-meter antennas in Australia to form a radio baseline of international dimensions. With a baseline of this size, a hitherto unprecedented resolving power of 2 milliarcseconds became available, and the arrangement was used for several years thereafter for the determination of extragalactic radio sources.

An improved version of this idea, known as the Tidbinbilla Interferometer, employed two antennas (64-meter and 26-meter) at the same site (Tidbinbilla) to form a short-baseline interferometer. With the advantage of real-time operation, the Tidbinbilla Interferometer started observing in 1980 and for many years thereafter engaged in a detailed survey of radio sources in the Southern Hemisphere.

In later years the VLBI concept for radio astronomy was extended to make use of an Earth-orbiting spacecraft as one element of an interferometer, with a large ground-based antenna as the other. The enormous increase in baseline length thus obtained—several Earth diameters—provided a proportionate increase in resolving power. Known as Orbiting VLBI (OVLBI), this complex experiment involved the DSN 64-meter antenna at Canberra, a new 64-meter antenna built by the Japanese space agency ISAS at Usuda, and NASA's geosynchronous Tracking and Data Relay Satellite (TDRS). The power of OVLBI was first demonstrated successfully in 1987 when experimenters detected extragalactic radio sources at 23 GHz using a baseline length of 2.15 Earth diameters. The success of this experiment stimulated widespread interest in the space-based VLBI technique. By 1997, NASA had become involved in a cooperative space-based VLBI program with Japan's ISAS and Russia's ASC. As its contribution to the international Space VLBI Project, NASA added to the DSN a network of 11-meter antennas operating in the higher frequency, more efficient Ka-band, and committed the 70-meter antenna to participate as a co-observatory with many other radio astronomy observatories around the world.

Like its radar-based twin, radio astronomy became a permanent element of the overall facility known worldwide as the Deep Space Network, not only because of the prestige that its distinguished investigators brought to NASA, but also because of the synergistic relationship that developed between the technologies required for deep space communications and those required for deep space science. It could be said, in a general sense, that what was good for radar and radio astronomy was also good for the DSN, and vice versa. But in a most specific way that sentiment applied to the 70-meter antennas.[2]

None of these initiatives would have been possible without the unique deep space connection that was provided by the Big Dishes of the NASA/JPL Deep Space Network. This is their story, and that of the unique group of men that conceived and created them.

It all began near a small, dry lake in a remote corner of California's Mojave High Desert.

High Desert

Railroad Town

Goldstone is not so much a place as a location, somewhere in the high Mojave Desert, near Goldstone Dry Lake. It doesn't have a name, it has geodetic coordinates: latitude 35.4° north, longitude 116.8° west. Barstow is a place; it has a real name: "City of Barstow, Pop. 23,000, Elev. 2100 ft." It is also in the high Mojave Desert of southern California. That it is near the location of Goldstone is incidental to this story, but the later history of Goldstone is inextricably linked to the early history of Barstow and the Mojave Desert.

In its historical publication *Once Upon a Desert,* the Mojave River Valley Museum traces the written history of Barstow back to the eighteenth-century explorations of a Spanish missionary, Fray Francisco Garces (Keeling 1976).

When Fray Garces traveled through this general area in 1776, he was searching for a practical route for immigration and trading that would connect the outposts of southern Arizona and New Mexico to the northernmost link in the chain of Spanish missions that, at that time, extended from the southern part of California down into Central America.

Then, as now, he would have gazed upon an endless landscape of red-brown baked earth and rocks, contoured only by low, weathered ridges that separated flat, or gently sloping, areas of sand and rocky debris. Here grew the desert plants, mostly small, tangled, scrubby bushes that matured each spring among the wildflowers, only to dry up each summer and blow away in the ever-present wind. Overhead the sky arched blue and cloudless to the horizon. Where the sky met the earth, the colors merged into a dusty haze that obscured the dividing line between them. Here and there a dried-up lakebed indicated a drainage area for winter rains and snow, or for the thunderstorm when flash floods could suddenly come roaring out of the surrounding canyons. Occasionally, jumbled piles of enormous rocks, sculpted by the wind and the heating and cooling effects of the sun and frost into fantastic shapes, thrust upward though the desert floor. And all around, the slightly comical shape of the Joshua trees, standing immutable and sullen among the rocks, relieved the monotony of the landscape.

It might not have surprised Fray Garces that the area through which he passed that day in 1776 would, by 1847, become the Old Spanish Trail carrying travelers of all types—raiders, slave traders, fur trappers, soldiers, explorers, stockmen, merchants, immigrants, gold seekers—between California and the eastern States, and that along this route, settlers would soon establish way stations to cater to the needs of the traveling public and their stock. Nor might he have been surprised that this barren land would hide mineral wealth of the exotic kind, gold and silver, and of the very mundane kind, borax, which in the years to follow would attract even more settlers to this forbidding place.

However, Fray Garces could not possibly have foreseen the dramatic changes that would take place in that area over the next hundred years, changes that would extend from the age of oxcarts and horse-drawn vehicles to the age of airplanes and space probes.

Following close upon the great western migrations of the 1840s and the gold rush fever of the 1850s, a frenzy of railroad building gripped the country in the 1860s. In Washington, the passage of the Homestead Law and the Railroad and other Land Grants bills in 1862 launched the construction of railroads heading west. The western half of the first transcontinental railroad would be built by the California-based Central Pacific Company under the powerful directorship of Huntington, Hopkins, Crocker, and Stanford. Pushing east from Sacramento, their line would cross the Sierra Nevada via the 7,000-foot-high Donner Pass and continue 690 miles east to meet the Union Pacific line driving west from St. Louis. The two opposing tracks met at a desolate point on the northern edge of the Great Salt Lake named Promontory, Utah, on May 10, 1869. For the first time, the Atlantic and Pacific coasts were connected by rail. The age of the Pony Express and the wagon train passed into history.

While it had been principally engaged with the challenging engineering and construction problems associated with building a railroad across the Sierra Nevada, the Central Pacific Railroad Company had also embarked upon a railroad building program that would soon give it a monopoly of all railroad business in California itself. Twenty years earlier, long before the completion of the transcontinental line, the Big Four had formed the Southern Pacific Railroad Company. Now the company planned to build a railroad and telegraph line connecting San Francisco with the Southern California communities as far south as San Diego.

When completed in 1876, the Southern Pacific route from Sacramento to Los Angeles ran roughly north-south through the San Joachin Valley along the western foothills of the High Sierra that forms the almost impenetrable backbone of California. Small farming, ranching, and mining communities soon established themselves along this route. One of them, Mojave, served the borax mines of Death Valley.

In 1882 Mojave became the junction point for the start of a new Southern Pacific line that was to meet the Atchison, Topeka, and Santa Fe Railway line from Albuquerque, New Mexico, at Needles on the Colorado River, near the point where the states of Nevada, Arizona, and California meet. Along the way, this line would pass through the thriving silver mining community of Waterman.

The completion of the Southern Pacific track from Mojave to Needles in 1883 gave California two railroad routes to the burgeoning cities of the East Coast: a northerly route through Sacramento and the high passes of the Sierra, and a southerly route via Mojave, Waterman, and the great deserts of the Southwest. Waterman, already established as a thriving community along the old coaching trail to the East, and no longer dependent upon bullock and mule teams to move its materials and supplies to Mojave or San Bernardino, rapidly expanded to become a railroad town.

For the Santa Fe railroad, the fierce competition with Southern Pacific for the lucrative Southern California rail business demanded a more efficient link to its transcontinental route to the East. This was accomplished in November 1885 when the company ran a line directly from San Diego to Waterman.

Everything, including its name, changed in Waterman when the new Santa Fe line was opened. Almost overnight, Waterman became an important rail junction on one of the two great routes across the continent. Huge switching yards were built. To serve the traveling public, a magnificent depot hotel and eating house was constructed on the south side of the Mojave River, and the name of the town was changed to Barstow to honor William Barstow Strong, president of the Santa Fe railroad (Moon 1980).

In the short space of forty years, the Old Spanish Trail had morphed into the Santa Fe railroad, with the desert town of Barstow as its most important center.

It rapidly became the lifeline of the desert community, serving the traveling public just as the old trail had done since earliest times.

Over the following decades, the town was relocated several times as the Santa Fe expanded. The population fluctuated as mining interests flourished and then declined when the leads gave out or became unprofitable. Across the neighboring desert, numerous small claims were worked for a while before being abandoned by their owners for more dependable and lucrative work with the railroad. One such mining enterprise was located about thirty miles from Barstow near the dry salty bed of Goldstone Lake. For a time the Goldstone mines thrived to the point that a Goldstone Club was established in downtown Barstow to provide a hangout and shady source of liquor for its members during the Prohibition era. A few years later the minehead and a few shacks, decaying in the fierce sun and desert wind, were all that remained to mark the passing of those hopes and dreams. No one could have thought, then, what the future held for Goldstone.

Gradually diesel trains superseded the old steam trains on the railroads of the nation, and later still the automobile and the airplane replaced trains for the mass of the traveling public. Barstow changed from a railroad passenger service center to an important switching center for rail freight moving across the continent.

When the national highway system began to take shape in the 1930s, Barstow became an important stop for travelers on the famous transcontinental Route 66 between Chicago and Los Angeles. In the 1930s the lights of Barstow, California, were a welcome sight for weary automobile travelers from the dust bowls of Oklahoma following Route 66 in their old Fords and Chevys to start a new life in the Golden State. Barstow moved sleepily through the 1930s, slowly increasing in size through its association with the Santa Fe railroad and the business accruing from its location on Route 66, while providing a supply center for the occasional mining prospector.

In 1940 Barstow came to the attention of the government as a possible site for a military reservation. Remote, in an area of little agricultural value, and located on both rail and road arteries to the East, Barstow was well suited to serve the needs of the military. Despite the objections of local mining interests, in August 1940 President Roosevelt designated an area of approximately a thousand square miles lying about fifty miles northeast of the sleepy desert town of Barstow as a military reservation. Initially called the Mojave Anti-Aircraft Range, it was officially renamed Camp Irwin in 1942 after Major General LeRoy Irwin, a World War I field artillery commander.

During World War II Camp Irwin was used by General Patton and others as a desert training area for the U.S. Army. It was deactivated at the end of the war.

In the years following 1945 it was used by the Army for tank training and battle exercises associated with the Korean and Vietnam wars. Following a long period of deactivation when it was used as a training post by the National Guard, the facility was renamed Fort Irwin and returned to active status in July 1981 as the U.S. Army's National Training Center (Morrison 2004).

As they did during World War II, Korea, Vietnam, and Desert Storm, soldiers of the U.S. Army and its allies, intent on their battle simulations and war-game exercises, continued to storm the heights and ambush the unseen enemy in their tanks and armored vehicles, while the immutable Joshua trees kept their silent watch across the unchanging desert.

A Small Dry Lake

The June day in 1956 when John Froehlich and his three JPL engineers drove from Pasadena to Barstow would have dawned clear and sunny as they left San Bernardino behind and slowly climbed the steep grade through the Cajon Pass up to the high desert and on toward Barstow. Along State Highway 15, the former U.S. Route 66, the Mojave Desert would have been in full bloom. The last of the colorful spring wildflowers would be holding court among the dour stick-figure Joshua trees that, unperturbed and unchanging like the desert itself, populated their surroundings. The air up there would have been very cool; as yet there was no smog and you could see "forever." But the desert would warm up by midday, and the intensity of the sun would indicate that summer was not far away. Altogether, it was the best time of year to visit the high desert.

Their destination was not really Barstow but the U.S. Army Headquarters at Camp Irwin, the training facility thirty-five miles beyond Barstow. There they were to meet with Army representatives to discuss JPL's search for a suitable site for testing its large developmental rocket engines. On the vast 640,000-acre property—remote, unoccupied, government-owned land where safety and noise considerations would be minimal—the Army would be willing, JPL believed, to find a small area where JPL could pursue its rocket-engine test program with least cost to the government and little impact on the Army's training operation.

The meetings were not productive, and over the next year JPL turned its attention to evaluating other possible sites for its new rocket-motor test facility.

Suddenly and quite unexpectedly, in January 1958, the Army informed JPL that it was no longer using Camp Irwin as a tank training site. Instead the National Guard would use the camp for a few summer months for its much more limited training exercises. By that time each of JPL's alternative sites had shown potential problems, and JPL interest in the Camp Irwin site revived. Toward the end of March, a JPL team was back at Camp Irwin investigating three areas, including a small dry lake named Goldstone, as possible sites for the rocket-test facility. A week later JPL's director, Dr. William H. Pickering, informed the commander of Camp Irwin: "While any one of the three suggested sites would be adaptable to our uses, and each one has some particular advantage, the Goldstone Lake area seems to be the most desirable, based on overall considerations" (Pickering 1958).

What Pickering did not say, though it might well have been in his mind, was that "overall considerations" then included much more than just the rocket-motor test facility that had been the subject of the original inquiry. For in the few short months since Froehlich's first meeting with the Army at Camp Irwin, the world had changed forever in a way that would catapult the word "Goldstone" onto the world stage as a vital link in the new age of space exploration.

To understand how that happened, let us step back somewhat to examine what else, in addition to jet propulsion testing, was going on at Jet Propulsion Laboratory in the years 1957–58.

Describing the origins of JPL in his 1977 history of Project Ranger, NASA's first lunar landing program, R. Cargill Hall wrote:

> Begun in 1936 under the auspices of the Guggenheim Aeronautical Laboratory of the California Institute of Technology, it had originated as a student rocket research project when the scientific community generally regarded rockets as an indulgence best left to students. In 1940 the Caltech rocket experimenters acquired an Army Air Corps contract and built facilities in northwestern Pasadena, at the foot of the San Gabriel Mountains in the Arroyo Seco wash. There they developed the first solid- and liquid-propellant rocket motors for jet-assisted takeoff of military aircraft. The enterprise was reorganized and renamed the Jet Propulsion Laboratory when, in 1944, after the advent of the German V-2 rocket, U.S. Army Ordnance awarded Caltech a contract to develop tactical ballistic missiles.
>
> Continuing to work for the Army into the 1950s, JPL engineers and scientists designed and developed the liquid-propellant WAC Corporal sounding rocket, the Corporal tactical missile, and the solid-propellant Sergeant tactical missile system. The Laboratory also pioneered in the development of radio telemetry and of various radio and inertial guidance systems for the Army's Redstone rocket arsenal in Huntsville, Alabama, where the director of research was Wernher von Braun. All the while, JPL, whose facilities were owned by the government, remained an Army establishment under the contract management of Caltech. Its posture and atmosphere were free-wheeling, academic, and innovative. By

> 1957 [when Froehlich first visited Camp Irwin] JPL Director Pickering . . . presided over a considerable laboratory complex nestled in the Arroyo Seco and populated by some 2,000 employees.

Appointed director of JPL in 1954 by Caltech president Lee DuBridge, Pickering lost no time in turning his preference for applied engineering and research and development into a standing policy for the laboratory. Thus in October 1957, when the Soviets surprised the world with *Sputnik*, the world's first earth-orbiting satellite, Pickering and his engineers and scientists were well prepared to collaborate with Wernher von Braun and his people at Huntsville in formulating the nation's immediate response to the Soviet challenge. For the orbiting of the United States' *Explorer 1* satellite in January 1958, JPL provided the solid-propellant upper stages for the Redstone launch vehicle provided by von Braun, the space-to-ground communications equipment, and the instrumentation for the satellite, which included the radiation monitoring experiment of James Van Allen.

Notwithstanding this obvious success, Pickering strongly believed that the United States should regain its stature in the eyes of the world by demonstrating a significant technological advance over the Soviets in rocketry and space flight. Not surprisingly, JPL already had a proposal in hand to further such an initiative, a proposal for a series of nine (unmanned) rocket flights to the Moon. The JPL proposal showed how the United States could use the technology then available to launch a simple spin-stabilized vehicle, similar in design to the Explorer satellite, to the vicinity of the Moon at short notice, possibly as early as June 1958. Initially, the Pickering and DuBridge proposal code-named Red Socks did not interest the Department of Defense, but in early 1958 it came to the attention of a new DoD agency called ARPA—the Advanced Research Projects Agency. At that time ARPA's new director, Roy Johnson, was responsible for the direction of all United States space projects. Eager "to surpass the Soviet Union in any way possible," Johnson believed an unmanned mission to the Moon would be a promising approach to beating the Russians in space, and he immediately carried the idea forward within the new agency (R. Cargill Hall 1977).

JPL did not have to wait long to see some action. On March 27, 1958, with President Eisenhower's approval, the secretary of defense announced that ARPA's space program would advance space flight technology and "determine our capability of exploring space in the vicinity of the moon, to obtain useful data concerning the moon, and to provide a close look at the moon." The program would be conducted as part of the United States' contribution to the International Geophysical Year. It would be generally known as the Pioneer program and was to consist of five launch opportunities—three conducted by the Air

Force, followed by two conducted by the Army. Space Technology Laboratories in Los Angeles would be the contractor for the Air Force, and Jet Propulsion Laboratory would be the contractor for the Army launches. Each contractor would be responsible for the design of its probe and instrumentation, the upper stages of its launch vehicle, and the ground-based tracking and data acquisition facilities. ARPA directed the Air Force to launch its lunar probes as soon as possible consistent with obtaining a useful amount of lunar data. The Army/JPL launches were planned for December 1958 (R. Cargill Hall 1977).

The problems of tracking the earlier Explorer flights once they achieved Earth orbit had amply demonstrated the need for a worldwide network of tracking stations, and ARPA authorized JPL to set up such a network. For many reasons, some obvious and some very subtle, the key station of the new network would be located at Goldstone and would have to be ready to support the Army's first lunar probe, scheduled for launch in December 1958 (Koppes 1982).

This, then, was the bigger picture prevailing at JPL when Pickering indicated to Brigadier General Jensen that "the Goldstone Lake area seems to be the most desirable, based on overall considerations." Suddenly, a small dry lake in the heart of the Mojave Desert had become a cornerstone of the United States space program.

Engineers at Pasadena immediately set about building their Pioneer lunar probes and an antenna to track them with, while in the USSR engineers and scientists were preparing their Luna probes for the first Russian attempts to place a probe on the surface of the Moon.

It was March 1958, and the race to the Moon was on.

3

World Network

Seen from any point on the Earth's surface, a deep space probe appears to move across the sky from east to west in much the same manner as any of the heavenly bodies. It does, in fact, become an addition to the solar system, albeit of shorter life and different orbit about the Sun than any of the natural objects comprising the solar system. Owing to the daily rotation of the Earth, the spacecraft appears to rise in the east, travel across the observer's field of view at what astronomers call sidereal rate (the same angular speed as, for example, the Sun), and set slowly in the west. Obviously, three such observers located at intervals of approximately 120 degrees of longitude around the Earth's surface would, between them, have a continuous view of a passing spacecraft.

This idea was the basis for the Advanced Research Projects Agency's concept of a World Network consisting of three tracking stations, equally spaced in longitude and each connected to a centrally located control center. Such a network would permit engineers and scientists at the control center to maintain continuous contact with a planetary spacecraft. As the Earth rotated, the radio links connecting the spacecraft to Earth would be passed sequentially from station to station, and the data and control signals carried on the radio links would pass in an unbroken stream between the spacecraft and the control center.

In March 1958, ARPA assigned the task of bringing this concept into being to JPL. Led by one of director William Pickering's brilliant young engineering managers named Eberhardt Rechtin, JPL had been developing the technology necessary to support this idea for some time, so that when ARPA gave the formal order to proceed, Rechtin and his team in the Telecommunications Division at JPL were ready to go into action.

Some of the technology, particularly the "phase-lock" Doppler receivers and telemetry data processing, was adaptable from the Microlock system that JPL had been using for their missile tracking work for the Army at White Sands Proving Ground in New Mexico. It was, in fact, the Microlock system that had provided the sparse tracking support for the Explorers. The new lunar probes called Pioneer would operate on a different and much higher frequency, 960 MHz versus 108 MHz, and would require much larger, and quite different, receiving antennas. However, the design, procurement, and erection of a suitable antenna in the brief period of nine months between the March go-ahead from ARPA and the Army's Pioneer launch date in December was recognized as the major challenge to JPL's ability to meet its newly assigned responsibility. Rechtin turned to Walter K. Victor and Robertson Stevens, two of his lead engineers, to coordinate the overall design and construction of the station. They in turn assigned the formidable task of implementing the new antenna to a smart, energetic young engineer named Bill Merrick.

Merrick first came to work for Bob Stevens in 1951 when JPL was heavily involved in its Corporal missile development contracts with the U.S. Army at White Sands Proving Ground. Clayton Koppes (1982) describes the prevailing environment at White Sands around that time:

> By 1950 White Sands boasted such comforts of permanence as a swimming pool and handsome quarters. Testing new and expensive things under a demanding timetable meant hard work and tension; relief was found in hard play. All-night poker games quickly became a staple of White Sands weeks. . . . Visits to Juárez took the chill off desert evenings. Missiles and Mexico made White Sands perhaps the ultimate in machismo for JPL males. . . . Away from home distractions, concentrating on work in a tense environment, playing hard, the two dozen or so top JPL officials formed an elite communications group that tended to cut across formal lines in the laboratory.

It was a challenging atmosphere of excitement and clearly defined goals, where successful missile flights were the obvious and ample reward to the inquiring minds of the engineers involved. Leading-edge technology in an atmosphere of secrecy, and a can-do approach to obstacles that stood in the way of that success, were the order of the day. In October 1951, Bill Merrick joined JPL as an engineer to work with the missile guidance, tracking, and telemetry group at White Sands. With his experience and unbounded enthusiasm for everything he undertook, it would be a perfect match.

For the next few years, 1951–54, Merrick worked as a key member of the Microlock Team at White Sands. Later, when JPL developed a radio guidance system operating at a frequency of around 10,000 MHz for the U.S. Army's Sergeant missiles, it was Bill Merrick who designed and built the prototype missile-tracking antenna. Thus, when ARPA called on JPL in 1958 to build a network of tracking stations in a hurry to meet the requirements of the Army's first lunar probe, the assignment for the antenna task naturally fell to Bill Merrick.

Given the limited time available and the fact that the design and fabrication of radio astronomy antennas normally took anywhere from eighteen months to seven years, it was obvious to Merrick that the JPL requirements could be met only by "minor modifications of an existing design." Merrick, by then head of the Antenna Structures and Optics Group, began an immediate search for potential suppliers.

An exhausting ten-day research trip to check out potential suppliers clearly pointed to a vendor of choice: the Blaw-Knox Company of Pittsburgh, Pennsylvania. There was, however, a major drawback to the Blaw-Knox antenna. It had been designed with a sidereal drive system for its intended use in radio astronomy applications rather than the automatic two-axis (hour angle and declination) tracking required for deep space probes. In all other respects, the antenna was quite adaptable for Merrick's purpose.

In 1999 Fred McLaughlin, one of the junior engineers working at Blaw-Knox at the time, recalled that January 1958 visit of the JPL team:

> There we were, working away at our drafting boards, and in walked Bill Merrick. Apparently they had been out doing an industry search for a large antenna. In his rather abrupt manner Merrick said, "Can you adapt these radio telescope antennas for space probe tracking?" In a hastily convened meeting at which our chief engineer asked, "Can we do that?" our immediate response was that we didn't know. Then we went quickly to the [drawing] boards to look at the acceleration requirements and the gearing and that sort of thing and the answer was "Yes, we can do that." Merrick then offered to send us ten thousand dollars for an immediate study, and within a couple of days JPL gave us authorization to move ahead.

That was typical of the way Merrick worked. By April, Merrick had the answer he was looking for. Now, confident that the structural design of this sidereal-mounted antenna would allow them to substitute more powerful dual hydraulic drives on each axis to satisfy the JPL requirements, and sensitive to the antenna's low cost and expected manufacturing precision, Merrick advised JPL to proceed with the contract. Acting on behalf of ARPA, JPL then authorized Blaw-Knox to build not one but three antennas with the specifications that had been negotiated, including delivery of the first antenna within six months.

While Merrick and his team were busy with the contract for the first antenna, Robertson Stevens, chief of JPL's Guidance Research Section, had his engineers consider two remaining key issues: the choice of operating frequency for the deep space radio link between probe and Earth, and the choice of a site for the first tracking station. Both issues needed to be resolved immediately so that work could begin on design of the receivers and new antenna drive systems and preparation of the site for the antenna.

Communications engineers at JPL were well aware that the growth potential of any deep space communications system would be limited by radio interference and noise at frequencies below about 500 MHz. The earlier Microlock tracking system had operated satisfactorily at 108 MHz for missile tracking and for the Earth orbiting Explorers, because the distances were relatively short and the signal received on the ground station antennas was therefore relatively strong. But now they were considering communications from deep space, where the strength of the signal received on the ground would be weaker, perhaps many thousands of times weaker. To be able to receive such signals clearly, and undisturbed by man-made electrical noise or radio noise from cosmic sources, a frequency around 1000 MHz or higher and a big antenna in a remote location were essential. With this very much in mind, but tempered somewhat by the state of available technology for very-low-noise receiving systems at the time, the decision was made to design the probes and the tracking stations to operate at a frequency of 960 MHz.

JPL's communications engineers began searching for a site for the first tracking station that would satisfy several important criteria. They were looking for a natural bowl, an area where the surrounding terrain would shield the antenna from nearby towns and passing vehicles. Areas with power lines, radio stations, radar transmitters, and passing aircraft were to be avoided. The underlying soil needed to be stable and of sufficient strength to support the foundations for a large tracking antenna, and the surrounding terrain should be suitable for the construction of access roads capable of carrying the heavy steel components for the antenna and its pedestal. A nearby town that could provide housing and a social support structure for the workers was also desirable. Finally, funding and time constraints mandated the use of existing government-owned property.

Goldstone met all of the site criteria, and the decision was made: Goldstone would be the site for the first antenna. On the same day that the Army and Air Force lunar probes were publicly authorized by ARPA, March 27, 1958, the director of JPL announced that JPL had selected the Goldstone Dry Lake area of the Camp Irwin Military Reservation as the site for the huge new antenna to track the Army's Pioneer lunar probe.

The space age had come to Barstow.

Stevens's engineers moved rapidly ahead with the modifications needed to convert the Blaw-Knox radio astronomy antenna to a precision-pointed spacecraft-tracking antenna. When completed, the new antenna would always keep its needle-width radio beam pointing at a distant target and relay its pointing direction to computers at JPL for orbit determination purposes. Sensitive receivers would convert the received telemetry signal to text, graphical, or numerical form or images for scientific data assessment.

Merrick and his team pressed rapidly forward with contracts for the construction of buildings and facilities and a road across the desert. In quick succession Merrick's planning brought forth buildings to house the receiving, data processing, and communications equipment, the hydraulic pumps for the antenna drive-motors, and the four diesel electric-power generators. Almost overnight, water and fuel storage tanks appeared on-site, followed closely by twelve large house trailers to be used as office, laboratory, and sleeping accommodation.

Blaw-Knox made good on its delivery guarantee, and on August 16, under the watchful eyes of Merrick's men, ironworkers began erecting the steel antenna structure. It was completed in mid-October. The final milestones, optical and beam-alignment tests followed by radio performance and tracking tests, represented the ultimate measure of success for this most ambitious and high-risk schedule.

Amid a flurry of last-minute details the station was pronounced "ready for launch" on December 3, 1958 (see figure 3.1). The new tracking station was to face its first challenge.

Three days later the U.S. Army launched its first probe to the Moon. It was called *Pioneer 3,* and it was launched not under the auspices of its parent organization, ARPA, but under the oversight of a brand-new organization called NASA.

Despite its successful launch on December 6, 1958, *Pioneer 3* failed to reach the Moon. Approximately twenty-four hours after leaving the launch pad in Florida, it fell back to Earth somewhere in Africa after reaching an altitude of only 63,500 miles. Between them, the small station in Puerto Rico and the large antenna at Goldstone tracked the probe and received the telemetry flight data throughout the entire flight. Goldstone had proved its worth.

Disappointing though *Pioneer 3* was, success came not long after. Launched three months later on March 3, 1959, *Pioneer 4* became the first U.S. spacecraft to leave Earth's gravitational field. Goldstone acquired the downlink signal

Figure 3.1. Pioneer Tracking Station near Goldstone, California, December 1958. NASA/JPL.

from the probe as soon as it came into view about six and a half hours after launch at a distance of some 60,000 miles from Earth. For the next ten hours, and for each of the two subsequent passes overhead, the new station vindicated its design by tracking *Pioneer 4* and receiving the downlink flight data. The station was still receiving a strong signal at a distance of 435,000 miles when the spacecraft batteries ran down and *Pioneer 4* fell silent (Corliss 1976).

The Army lunar-probe program ended with *Pioneer 4*, but for the Goldstone station, life was just beginning. Named Goldstone Pioneer station to commemorate these historic events, the station served NASA as a key element in the Deep Space Network until 1981 when it was decommissioned. Four years later, in December 1985, the U.S. Department of the Interior designated the Pioneer site as a National Historic Landmark (National Park Service) in recognition of its key significance in the formation of the U.S. space program.

Through most of 1958, while the attention of most of JPL had been focused on the exciting events that were taking place at Cape Canaveral in Florida and at Goldstone in the Mojave, another event, of even greater significance to JPL, was taking place in Washington, D.C.

Early that year, even as it assigned responsibility for tracking the Army Pioneers to JPL, ARPA was planning to expand its tracking facilities into a truly worldwide network for supporting the exploration of deep space. In ARPA terms, it was to be called the World Net. In addition to the station then under construction at Goldstone, ARPA's plans called for two overseas stations, one at Luzon in the Philippines, the other in Nigeria. These plans were strongly influenced by ideas from Rechtin's group at JPL and were based solely on requirements for optimum tracking coverage of deep space probes. In July of 1958 the Department of Defense questioned the ARPA plan's suitability for all planned U.S. missions. The reassessment that followed showed that a more northerly site in Spain and a more southerly site in Australia would indeed offer some advantages, particularly for covering the orbits of possible future manned space flights in addition to those of deep space probes and Earth-orbiting satellites. As it turned out, however, ARPA never got the opportunity to build its World Net.

Ever since early 1958 when President Eisenhower sent Congress his proposal to establish a new and independent agency to oversee the nation's civilian space program, JPL had been considering its options. It had begun to find the idea of a civilian, as distinct from a military, affiliation increasingly attractive. One of JPL's top officials observed that JPL and Caltech had "almost . . . a moral obligation to see that the assets in a unique organization like JPL . . . are not restricted to serve only the military." On July 29, 1958, the president signed the bill that created the National Aeronautics and Space Administration. It became effective on October 1, and JPL's transfer from the Army followed shortly thereafter on December 3. Three days later JPL's new station at Goldstone, operating now under NASA ownership, tracked its first space probe, *Pioneer 3,* toward the Moon (Koppes 1982).

The formation of NASA did not create or lead to any break in the nation's space program. For JPL it was primarily a transfer of authority and funding. Instead of an Army contract, JPL now worked under a NASA contract. On the other hand, NASA acquired all of JPL's experienced personnel, technology, and facilities including Goldstone and Rechtin's vision of a worldwide network for tracking deep space probes. NASA's plans were in keeping with this vision, and the concept of the World Net survived the political upheavals of 1958 intact (Corliss 1976).

The rush of events that followed the establishment of NASA was described by historian William Corliss (1976):

> The Army Explorers and early Pioneers had been rushed into hardware more in response to Russian and interservice rivalries than as parts of a long-term, carefully planned, national effort in space exploration. With NASA assuming responsibility for all nonmilitary space activities and with the international space race in full swing, ambitious plans were the order-of-the-day. The pressure of the requirements of the *Ranger, Surveyor* and *Mariner* missions then taking shape on NASA's planning boards added more momentum for the completion of what ARPA had called the World Net before NASA took over. To complete the World Net, NASA needed two more sites with 26 m (85 ft) diameter dishes roughly 120 degrees of longitude apart.

NASA had already inherited one of the three ARPA antennas that had been purchased under JPL's persuasion the previous year. That one was now in operation at Goldstone; the other two were awaiting shipment overseas. But it was no longer clear where should they go, or which agency should own them. In the rather involved negotiations that followed, NASA eventually obtained ownership of both antennas. It decided to put one in Australia, the other in South Africa.

The first overseas station of the NASA Deep Space Network was located at Woomera in the state of South Australia, approximately 110 degrees west of the longitude of Goldstone. Woomera was already an established missile test center for the British Commonwealth and met all the criteria that JPL had established for a deep space tracking station site. It too was in a desert region not unlike that of Goldstone, although even more remote from a major population center. There was also thought to be a ready supply of high-quality technical manpower available nearby as a consequence of the missile test activities.

Construction of the site and erection of the antenna followed pretty much the same course as at Goldstone. Construction began in March 1960 and was completed by September (see figure 3.2). After the necessary calibration and checkout and performance tests, the Australian technical staff took over and, with the help of a few key technical personnel from JPL, soon had the station in full operation.

To maintain continuous coverage of a departing planetary spacecraft, the third station in NASA's Deep Space Network had to be located, as we have seen, on or near the longitude of Spain or South Africa. Spain was the first choice, but

Figure 3.2. The 26-Meter Antenna at Island Lagoon near Woomera, South Australia, September 1960. NASA/JPL.

diplomatic difficulties associated with the proposed Spanish site led to a decision to build the station in South Africa. The South African government offered a site near the Hartebeestpoort Dam about forty miles north of Johannesburg. The site was intended for a government-owned radio research station and, although not in a desert region, it met all of the appropriate site criteria. With the same construction and logistics planning techniques used as at Woomera, the material was delivered to the site in time for construction to begin in January 1961. Rather like the Goldstone station, the Johannesburg station was constructed at a frantic pace to meet the launch schedule for a lunar spacecraft. This time it was *Ranger 1,* the first U.S. lunar lander spacecraft, designed and built at JPL for the NASA Ranger program. The station was completed, checked out, and put in operation in time for the first *Ranger 1* tests in July 1961 (see figure 3.3).

The original World Net, a vision created by JPL under ARPA authority, had now been brought to reality by JPL under NASA authority. It was the same vision; only the name had changed. In keeping, perhaps, with JPL's missile test range origins, where everything that didn't fly was an "instrumentation facility," the World Net in the real world assumed the somewhat prosaic title of Deep Space Instrumentation Facility, or DSIF.

It was 1961, and for the next forty years the DSIF, later to be called the NASA Deep Space Network, would form the basis for all the NASA deep space programs to explore the planets, the solar system, and beyond. Continuously expanded and improved, its operational functions orchestrated from a control center at JPL, the network would reach a standard of excellence unmatched by any other facility of its sort in the world. The first three stations of the worldwide DSIF, now in operation at Goldstone, Woomera, and Johannesburg, were just the beginning.

Figure 3.3. The 26-Meter Polar-Mount Antenna at Hartebeestpoort Dam near Johannesburg, South Africa, July 1961. NASA/JPL.

While JPL was busy with the Pioneer lunar probes as the forerunner to the more ambitious Ranger hard-lander missions to the Moon in 1959, NASA became interested in evaluating the use of large balloon-type structures as passive Earth-orbiting communications satellites. In the original concept, two AT&T engineers, John R. Pierce and Rudolf Kompfner, suggested that radio signals reflected off a large inflatable metallic balloon in Earth orbit at an altitude of about a thousand miles could provide two-way communications between widely separated stations during the time that the balloon was in their mutual view. NASA decided to support a joint experiment with AT&T to test the concept using transmitting and receiving stations on the East and West Coasts of the United States. For obvious reasons they named the experiment Project Echo. NASA's Langley Research Center would provide and launch the balloon, and JPL's new Pioneer station would be the West Coast terminal, while Bell Telephone Laboratories would provide the terminal facilities for the joint experiment on the East Coast (Butrica 1997).

Bell Labs had the necessary equipment to transmit an East-Coast-to-West-Coast link at 960 MHz and to receive a West-to-East signal at 2390 MHz. Goldstone would use its new 26-meter antenna (figure 3.4) to receive the signal from the East, but it had no capability to transmit the West-to-East signal via the Echo balloon to Bell Labs. Goldstone obviously needed a second antenna and needed it quickly. JPL called in Bill Merrick.

As Merrick put it, Walt Victor, his boss in the communications group at JPL, "wanted to find out how quickly you could build one of these things and how much one of these antennas could be bought for. So he asked me to design a real cheap one. This cost two hundred thousand [dollars] and was installed at what was called the Echo site at Goldstone in five months" (W. Merrick 1992).

Merrick and his team designed an antenna in short order, expedited procurement of the components, and began construction in July 1959. By the end of the year, Goldstone's second antenna was being equipped and calibrated to support the Echo balloon experiments. The Echo antenna differed from the Pioneer antenna in its higher-speed azimuth-elevation (Az-El) drive system and its state-of-the-art 10,000-watt microwave transmitter operating at 2390 MHz (S-band). Cheap it may have been, but short on performance it was not. Later Rechtin, Victor, Stevens, and others used that antenna to make JPL's first major success in planetary radar. In that now-famous experiment on March 10, 1961, radio signals transmitted from the Echo antenna and reflected by the planet Venus were unambiguously detected by special receiving equipment installed on the Pioneer antenna. The data that resulted from that experiment was of immense scientific importance. Among other things, it led to a hundredfold improve-

Figure 3.4. The 26-Meter Azimuth-Elevation-Mount Antenna at Goldstone Echo site, March 1962. NASA/JPL.

ment in knowledge of the value of the Astronomical Unit (AU), a basic parameter used by astronomers and a fundamental component of all spacecraft and planetary orbit determination processes (Mudgway 2001).

Although there were then two antennas at Goldstone, by March 1961 it was becoming apparent that a third antenna was needed to accommodate the tracking load of the many new deep-space missions appearing in the NASA long-range launch schedules.

Of immediate concern to mission planners at JPL were the seven new Ranger spacecraft then being built at JPL. They were intended to carry scientific experiment packages and television cameras to the Moon, where measurements and images of the surface would be transmitted to Earth in the final moments before impact. The data and pictures would be received by the Goldstone tracking station. But there was more cause for concern. In parallel with the Ranger program,

Figure 3.5. The 26-Meter Polar-Mount Antenna at Goldstone Echo Site, June 1962. NASA/JPL.

JPL was also planning its first foray into deep space. Engineers were busy designing the first of the Mariner spacecraft to make a close flyby of the planet Venus. The first Ranger launches were scheduled for August 1961, and just about one year later the first Mariner would blast off for Venus. There was no time to spare.

With the Ranger and Mariner programs running in parallel, NASA/JPL was determined to challenge the Soviets for superiority in space on two fronts, the Moon and the planets, beginning with the planet Venus.

Obviously, more missions in space required more antennas on Earth to keep track of them. Furthermore, JPL's communications engineers needed a separate antenna at Goldstone where they could develop and evaluate new ideas under field conditions. Both of these problems were solved in mid-1962. A new polar-mount antenna (figure 3.5) was built at the Echo site to support the first Mariner mission to Venus, and a few weeks later the former Az-El antenna was relocated to another site at Goldstone where it served for many years as a JPL field laboratory for communications research and development (Mudgway 2001).

Now, with two operational polar-mount antennas at Goldstone and a similar polar-mount antenna at each of the overseas stations in Australia and South Africa, the DSIF was ready to support the two Mariner missions to Venus planned for launch late in 1962. For the first time, the DSIF would be dealing with two deep-space missions simultaneously, the Ranger missions to the Moon and the Mariner missions to Venus. It would be fully immersed in multiple-mission operations, an environment that formed the principal driver for most of the subsequent expansion of the network.

So it was that by 1962 the concept of a world network had been realized and demonstrated with several live space missions. Owned by NASA and managed by JPL for Caltech, the DSIF had grown from a single 26-meter antenna at Goldstone to a worldwide network of four online antennas, connected by an intricate web of communications links with JPL's Network Operations Control Center in Pasadena. A supporting research and development facility had been established at Goldstone, and the successful Venus radar experiment had brought JPL and its deep space expertise to the attention of the world scientific community.

The year 1962 proved an eventful one for the deep space programs of both the United States and the Soviet Union. For various reasons none of JPL's first five Ranger lunar missions was successful, nor was the first Mariner attempt to reach Venus. However, the second Mariner spacecraft flew by Venus in December and, during a 42-minute scan of the planet, transmitted to Goldstone important data on Venus's atmosphere and surface features. It was considered an outstanding "first," a great success. JPL's director William Pickering appeared on the cover of *Time* magazine on March 8, 1963 ("Voyage to the Morning Star.") Referring to the first Mariner encounter with Venus, *Time*'s writers waxed eloquent:

> On that December day, though, the morning star held a special attraction for the men of Caltech's Jet Propulsion Laboratory. Almost as if they could see it all happening, they squinted into 36 million miles of space, out into the vicinity of Venus, where for the first time in history a man-made space traveler was cruising into range. A gold and gleaming machine, sporting angular purple wings and unblinking electronic eyes, was swooping toward its target. Mariner II was giving earthbound scientists their first close look at the distant planet that has tugged so long at their adventurous imagination. And when Mariner's radioed reports were finally decoded by the JPL crew that had built the spacecraft and sent it on its way, Venus would never seem quite the same again.

For the Russian program, the situation in 1961–62 was less encouraging. Of five attempts to reach Venus and three to reach Mars, only one had been, at best, partially successful (Siddiqi 2002).

Although the United States had gotten off to a slow start in the international space race, the NASA deep space program was rapidly catching up. It was about this time that JPL took another small step that presaged a giant leap forward in deep space communications technology: the Advanced Antenna System (AAS), with Bill Merrick as project manager.

4 Reaching Farther Out

Even as the DSIF struggled to keep pace with NASA's initial program to explore the surface of the Moon with robotic spacecraft, Dr. Eberhardt Rechtin's vision was reaching further out, far beyond the Moon, to Venus and Mars and, maybe, to planets even more distant. He had established the basic engineering technology for deep space communications and had brought the original concept of a World Net into being in the shape of the NASA Deep Space Instrumentation Facility.[1] The complexities of day-to-day operation of the worldwide network had been turned over to others whose expertise lay more in the area of routine, round-the-clock operations. With all of that in place, reaching for the Moon became history to Rechtin's mind, and his attention turned to the next step, reaching for the planets.

In November 1961 the American Society of Mechanical Engineers held its annual winter meeting in the city of New York. The agenda included the presentation of three papers on the general subject "Large Ground Antennas" (Rechtin, Rule, and Stevens 1962). Two of the presenters, Rechtin and Stevens, were high-level-management engineers from Jet Propulsion Laboratory, NASA's new space research facility in Pasadena. The third author, Bruce Rule, was chief engineer of the Mount Wilson and Mount Palomar observatories. All were from California, and all were deeply interested in the technical challenges associated with the design and construction of large antennas. By the standards of that time, "large antennas" meant antennas of a diameter of 200 feet (roughly 60 meters) or greater, which could be used for receiving very weak radio signals from space probes or natural radio sources in deep space.

Eb Rechtin was chief of the Telecommunications Division and program director of NASA/JPL's Deep Space Instrumentation Facility, the worldwide network of tracking stations that was already in operation and providing the communications links for NASA lunar probes. Within JPL he was informally known as the Father of the DSIF. Poised and very articulate, Rechtin began his ASME presentation by reviewing the capabilities of the current (1961) network and forecasting the enhancements that would be required to meet the ambitious programs of planetary exploration that NASA was planning for the next decade.

"Our ability to communicate with space probes depends both on the capabilities of the spacecraft and the capabilities of the ground station," he said. "The present capabilities, in combination with the Ranger and Mariner class of spacecraft, are capable of very good communications from the Moon, including the transmission of high-quality pictures every few seconds. They are also capable of sending quite useful data from the nearby planets, including good pictures every half hour or so. By using precision ranging and Doppler signals to and from the spacecraft, and by following the deep space probes for several days during their journey, the [present] facilities can determine the orbit of the probe with sufficient accuracy so that maneuvers can be made which will bring the probe to within a hundred kilometers of a designated point on the Moon, or to one planetary diameter of the planets" (Rechtin et al. 1962).

He pointed out that the existing 85-foot-diameter antennas were adequate for carrying out these kinds of tracking functions and were also capable of sending the commands necessary to control the robotic functions of the spacecraft, such as switching the science experiments or the propulsion or attitude control systems on and off. It was comparatively easy, and well within the capability of the technology of the early 1960s, to build bigger ground transmitters to enhance the uplink. But the capability of the downlinks, the radio link between spacecraft and Earth antennas that carried the science data, was limited by the power, size, and weight of the spacecraft. Bigger antennas and transmitters on the spacecraft would require larger and heavier spacecraft, and therefore more powerful rockets to boost them into space. However, in terms of launch weight and spacecraft trajectories to the planets, even the nearest ones, every ounce of additional spacecraft weight translated into an enormous increase in launching-rocket performance. That was the key issue, the trade-off that could not be made at that time, as Rechtin made clear: "The principal need for larger antennas is therefore to increase the communication [capability] from the spacecraft to the Earth." With the improvements he proposed, the downlink performance could be raised to a level "corresponding to real-time television from lunar orbiters and lunar landers; good pictures of the nearby planets taken from orbiters that remain with the planets throughout their full journey round the Sun; and useful scientific data even from the edge of the solar system." The audience gasped—live television from the Moon, pictures from Venus and Mars, science data from the edge of the solar system? In 1961 the world was only just getting used to live television from local sporting events, and man-made objects orbiting the Earth were still regarded as a demonstration of superior superpower technology in the Cold War, not as objects for scientific or commercial use.

"To accomplish these higher goals," said Rechtin, "we will add three antennas[2] of approximately sixty-four meters [210 feet] diameter to our present twenty-six-meter [85-foot] antennas.[3] Particularly important for the antenna designers will be an increase in both the listening and transmitting frequencies, a reduction in system noise temperature and an increase in the amount of power transmitted by our antennas."

With typical Rechtin lucidity, the objectives had been set. It would be up to his engineers to design an antenna to meet those objectives. He wanted an antenna that not only would be almost three times greater in diameter but also would have more sensitive radio hearing ability (lower system noise temperature) and would speak with a louder radio voice (increased transmitter power). It appeared that he wanted to have his cosmic cake, as it were, and eat it too.

It was no accident of convenience that Rechtin and his two colleagues were addressing a meeting of mechanical engineers that evening, for the subject of their discussion would be essentially a mechanical engineering enterprise of unparalleled complexity. "In brief," Rechtin concluded, "what is needed are antennas of between two hundred and two hundred fifty feet in diameter, with a surface tolerance of approximately one quarter inch, capable of being pointed to a few hundredths of a degree under [normal] weather conditions, at all three DSIF sites, costing between ten and twelve million dollars [each], and being first available in California by January 1, 1965."

He wrapped up his presentation by alluding to the difficulties of meeting the critical specifications of such a large antenna for deep space communications and noting some of the obvious design conflicts that would be encountered—like those between the radio communicators, who would prefer the largest-diameter antenna they could get, and the mechanical engineers, who realized that their design problems increased drastically as the diameter increased. There were also the instrumentation designers, who wished to locate the angle readout and deflection sensors at critical points in the structure but were preempted by the design requirements of the servomechanical and structural engineers. Finally, there were the enthusiastic antenna designers who wanted to build a very advanced antenna of "world prestige" diameter, but who faced reality in the form of the deep space communications customers, the people who would actually use it to support the future deep space missions. They needed the antenna by 1965, four years hence, and at a fixed and competitive cost.

Before turning the podium over to Bob Stevens, Rechtin filled in some background and provided a snapshot of the current status of the project. Acting as NASA's agent for the development of deep space communications, JPL had been studying the problem of building very large antennas for some time. It had been assisted in the feasibility studies by four special groups formed from some of the nation's most qualified industrial contractors; Blaw-Knox/Dalmo-Victor/Alpha Corporation, Hughes Aircraft/Consolidated Western Steel, North American Aviation, and Westinghouse Electric Corporation. These studies had determined that the concept of a single steerable paraboloidal antenna with a diameter of 200 to 250 feet was of sufficient merit to warrant further, more detailed study. It had also become apparent that the three most critical features of the design would be the antenna mounting system with its the azimuth and elevation bearings, the instrumentation to indicate the direction in which the antenna was pointing, and the servomechanisms to actually point the antenna in the desired direction with sufficient accuracy. The extent to which the design specifications could be realized in each of these areas would, said Rechtin, make or break the project. However, after a detailed evaluation of the technical material from the feasibility studies, NASA/JPL had decided to proceed with a preliminary detailed design. This work would be carried out by Blaw-Knox on a schedule that would, by mid-1962, allow JPL to call for construction bids leading to a construction contract by about September 1962.

Speculating on whether or not all this would actually take place, Rechtin observed that it depended "on the present studies, and on the decision of our government to proceed further with this element of the exploration of space." What Rechtin also knew, but did not say, was that it also depended on the outcome of the ambitious Ranger and Mariner/Venus missions then in the advanced planning stages at JPL.

Within JPL's matrix-type organization of the time, Robertson Stevens led a section of the large Telecommunications Division that was charged with developing new techniques and technologies for deep space communications. It was called the Communications Elements Research Section, and it included one unit devoted to antennas and optics. The Telecommunications Division contained several sections similar to Stevens's, each devoted to a specific aspect of deep space communications; two of these were Communications Systems, led by Walter K. Victor, and Communications Engineering and Operations, led by Nicholas A. Renzetti. Under the matrix organization scheme, Rechtin played a dual role as chief of the Telecommunications Division and also program director of the DSIF. In this latter role he represented JPL to NASA headquarters insofar as the tracking and data acquisition functions of the DSIF were concerned. He was the channel through which NASA funding found its way to support the activities of the DSIF, including its operation and, of course, the critical engineering and research work on which the DSIF depended for its continued development and expansion.

Many of the brilliant young engineers who largely comprised the Telecommunications Division had been with Rechtin, Victor, and Stevens since the early days of JPL when, as a military contractor, it had developed guided missile systems for the U.S. Army. Rechtin, Victor, and Stevens had been an integral part of the JPL Team working at White Sands, New Mexico, in the early 1950s, and they had been key participants in establishing the Microlock tracking stations, JPL's original satellite tracking network and the forerunner of the DSIF, in the late 1950s. With all that experience behind them, and with the existing network of 85-foot antennas now up and running, these three men were powerful advocates for the big new 210-foot antennas. Rechtin had spoken; Bob Stevens, Walt Victor's antenna expert, spoke next.

Bob Stevens was an affable, pipe-smoking man, somewhat laid-back, and rather shy at first acquaintance. He played golf and classical guitar with equal enthusiasm, and bicycled to and from the lab each day, five miles across the city of Pasadena. His gentle manner belied his depth of knowledge and the maturity of his experience. That day, he came straight to the point: "The application of two-hundred to two-hundred-fifty-foot-diameter, low-noise, precision antennas can increase the communications capability of the [present] ground stations in the NASA/JPL Deep Space Instrumentation Facility by a factor of about ten. Studies at JPL indicate that building and installing the large antennas is a practical and economic way of providing some of the increased communication needs of future spacecraft missions" (Rechtin et al. 1962).

As Stevens described them, the technical and operational requirements of the AAS were mind-boggling. They began with the influence of NASA's major space exploration missions on the AAS design.

It would take about five years to complete the first antenna and bring it into active service. The final design must therefore anticipate the future needs of planetary explorers and, once determined, the design must be "frozen" so that the future spacecraft might be designed to use the additional new capability that it would provide. Finally, it must be completed on time, because the launch and encounter dates of planetary spacecraft were determined, not by the whim of mankind, but by the intricate laws of motion that rule all members of our solar system including Earth. It became almost mandatory, therefore, that the ultimate design employ a conservative engineering approach, based on modest extrapolation from existing engineering practices and design methods.

With those constraints in mind, Stevens described three prime objectives that would be the basis for a construction contract for the first antenna at Goldstone: (1) to improve existing communications capability by a factor of about ten, (2) to design, fabricate, and erect the first antenna for a total cost of about $10 million, and (3) to complete the first antenna by late 1965.

In the esoteric world of deep space communications, the ability to detect very weak signals from a distant spacecraft is directly related to the effective receiving area of the antenna. Double the effective receiving area, and the antenna will collect twice as much power from a radio wave falling upon it, making a just-not-detectable signal into a just-detectable signal, or making a very weak signal just that much stronger and clearer. To achieve a tenfold increase over the performance of the existing 85-foot-diameter antennas, a new antenna with a tenfold increase in effective receiving area would be required, and the new antenna would have to be at least as efficient in collecting radio frequency as the present smaller antennas. This meant that the surface of the new antenna would have to have the same parabolic shape and the same accuracy of contour as the present antennas. Stevens's engineers had calculated that it would take a parabolic dish between 200 and 250 feet in diameter, whose reflecting surface contour would not deviate from the proper shape by more than a quarter inch under all foreseeable conditions, to do it. It would indeed be a Big Dish.

But there was more. The total signal energy collected by the Big Dish was only half the story. There was also the question of noise—that is, radio "noise," similar to the kind of noise heard on FM or seen on a TV screen between channels—to be considered. When dealing with really weak radio signals from space, it doesn't take much radio noise to smother the signal completely. It is, in fact, the ratio of the signal power to the noise power that determines the detectability of the signal. Communications engineers use this signal-to-noise ratio (SNR) as a basic measure of the performance of all communications systems. Halving the noise is as effective as doubling the signal—and, conversely, twice as much noise is as bad as half as much signal. So, Stevens explained, noise reduction was right up there with antenna diameter and shape as critical factors in the design of the Big Dish.

To appreciate the importance of the noise-reduction problem, it is necessary to understand the source of this radio noise. At the frequencies used by the DSIF to communicate with planetary spacecraft (2100 GHz to 2300 GHz at that time), the sensitivity of a radio receiver is no longer determined by the radio noise generated by the electronics in the receiver, as is the case for a standard FM or TV receiver. Instead, the unwanted radio noise is generated by the antenna itself, and by various external sources such as the cosmos, Earth's atmosphere, and the warm terrain surrounding the antenna structure. Communications engineers have a way of measuring this form of radiation using a scale graduated in units called kelvin (K). It is rather like the Celsius temperature scale, and is in fact called the radio noise-equivalent temperature scale, or noise temperature scale for short. For reasons too complex to discuss here, zero noise temperature on the Kelvin scale corresponds to physical "absolute" zero, or –273° Celsius. Using the Kelvin scale, then, radio noise can be expressed as a noise-temperature number, and these numbers can be added or subtracted arithmetically. For example, the noise temperature of a conventional microwave receiver would be about 1,000–2,000 K. For the type of maser receivers used by the DSIF, it would be about 10 K, while radiation picked up by the antenna from the warm ground would contribute 5–20 K, noise owing to Earth's atmosphere would add 3–10 K, and noise from the cosmos or galactic background would account for a further 1–5 K. Adding these all together shows that the combination of a low-noise maser receiver with a well-designed antenna can result in an overall receiving-system noise temperature in the range of 20–45 K.

If a system such as that could be realized, it would have even more than a tenfold improvement over the existing systems that used the 85-foot antennas. Of course, there would be some constraints on where such a system could be located to avoid overwhelming the antenna with man-made radio interference from radio, television, electrical machinery, commercial communications, radar, and so forth. An area remote from any industrial center would be mandatory. Furthermore, this would be no delicate laboratory instrument but a rugged piece of sophisticated machinery that would, except for brief routine maintenance periods, be operated continuously, in a remote and often hostile environment, in all weather conditions, from horizon to horizon, with no perceptible degradation of performance.

That was challenge enough, but Stevens hadn't finished. There was still the question of how to point this huge machine in the right direction.

A radio antenna of this type has a narrow, sharply defined radio beam, much as a searchlight or flashlight has a narrow light beam. The width of the beam is measured in degrees—one or two degrees for a searchlight, five to ten degrees for a flashlight. The beamwidth is a measure of the intensity, or resolution, of the beam and is a function of the size of the reflector: the bigger the reflector, the narrower the beam. Stevens's engineers had estimated that, when operating at the frequencies for which it was to be designed (2.3 GHz), the Big Dish with its diameter of 200–250 feet would produce a radio beam just a few hundredths of a degree wide. With a beam that narrow, the slightest error in pointing it at a target, especially one at the distance of Mars, would have a very significant effect on the strength of the signal received from the target spacecraft. The numbers showed that, to make it a viable operational instrument, the antenna would have to be pointed to a defined target with an accuracy of three hundredths of a degree or better. Not just sometimes, not after hours of calibration, but all the time, right now, in calm weather or fifty-mile-an-hour wind, in blazing sun, dense fog, driving rain, in ice or hail or snow, in the high desert of California, in the mountains of Spain, or on the bleak plateau of southeastern Australia. As Stevens warmed to his subject, the engineers assembled in New York began to appreciate the measure of the challenge.

There were two ways in which this might be done, and each had its advantages and disadvantages. Stevens outlined them both, to demonstrate the rationale that underlay the final antenna-pointing specifications.

In the first method, "absolute pointing mode," the antenna is instructed to point its beam to an absolute direction defined in some coordinate system that the antenna recognizes. The absolute direction would be the direction in space from which the spacecraft signal was expected to come. But how sure could you be that the absolute direction was not a bit off? There were a lot of uncertainties involved in computing that direction. Uncertainty in the true position of the spacecraft, uncertainty of the true position of the antenna location on the surface of the Earth, the bending effect of the Earth's ionosphere as the radio propagation path from the spacecraft passed through it, and, finally, errors in the actual direction of the antenna beam relative to the antenna's angular position as indicated by its internal instrumentation system. When this was all factored in, it appeared that with the "absolute" mode the beam could be pointed with confidence to within about three hundredths of a degree of the true direction of arrival of the spacecraft signal. That translated into a loss of about 20 percent of the power that would be expected if the beam could be pointed precisely in the correct direction of arrival of the signal. Twenty percent loss, that was significant. But that's what the numbers showed.

There was another technique for pointing an antenna to a spacecraft as it moved across the sky. This was the "angle-tracking mode." The DSIF used it regularly to drive the existing 85-foot antennas to follow their targets. A smaller antenna, with a very wide beam, pointed the larger, narrow-beam antenna in approximately the right direction so that it could lock on to the spacecraft signal. Thereafter it would track the spacecraft automatically. However, this technique required much more complex microwave equipment, with its consequent additional radio noise, than the more straightforward absolute-pointing arrangement. While the angle-tracking system was satisfactory for the 85-foot antennas, it was far from clear at that time that the 210-foot antenna could be provided with a satisfactory angle-tracking system—that is, one that would not inject so much noise into the receivers as to wipe out the advantages of the otherwise low-noise design. But, said Stevens, his microwave engineers at JPL were working on it. When the pointing errors introduced by the servo systems that actually drove the antenna were considered, it appeared that the angle-tracking mode would produce about twice the degradation of the desired signal as the absolute-pointing mode, but it would be proportionately easier to implement and to operate in the field. The conclusion was clear. Until they had an acceptable low-noise microwave design for an angle-tracking system, they would go with the absolute-pointing arrangement.

Stevens turned next to the weather environments in which the Big Dishes—all three of them, eventually—would have to operate continuously, without failure or degradation of performance. Obviously the specific weather environments depended on the locations of the three DSIF sites. It was an advantage that the existing sites were all in the Earth's temperate zone, about 30 degrees of latitude north and south, in California, Australia, and South Africa.[4]

Measurements of wind and weather data had been made at the existing sites since they had been in operation, but the time span—three to four years—was too short to give much confidence in micrometeorological data. This applied especially to the wind data, perhaps the most critical environmental parameter for a structure of the size of the Big Dish. The detailed picture of the wind conditions depended on the local topography of the exact site of each antenna, and these had not yet been decided, except in a general way for Goldstone. The data showed that the average wind velocity about thirty feet above the ground was less than 30 miles per hour 99 percent of the time. However, winds to 60 mph were not uncommon and occurred several times a year. Since this was all there was to go on, the specifications were written accordingly. The Big Dish must be capable of its full pointing accuracy in wind speeds up to 30 mph, and with accuracy degraded by a factor of not more than four at 60 mph wind velocity. It would be placed in a safe, stowed position when the wind reached 70 mph, and would survive winds up to 120 mph in that position without damage. The designers wanted to be very sure it would not simply blow over. There was also a specification for the allowable distortion from heating by the Sun, and for the loading by snow and ice.

Finally, the antennas would have to be erected in remote areas, fifty or even a hundred miles distant from the nearest towns, to avoid the adverse effect of man-made electrical noise. The contingent problems of access roads for the

Figure 4.1. Radio Telescope at Jodrell Bank, Manchester, England. The 250-foot-diameter steerable reflector at Jodrell Bank, University of Manchester, has a double-tower structure to support the elevation-axis bearings at the periphery of the reflector, and a very wide wheel-and-track base around which the towers rotate in azimuth. The elevation drive wheel and antenna surface backup structure are clearly seen. A slender column placed at the center of the antenna supports the feeds for the receivers at the point of focus. The control room and data handling equipment are located in nearby buildings, not seen in this photograph. NASA/JPL.

transportation of construction material and personnel and, later, commuting or accommodation for the personnel that would be required to operate and maintain the antenna would also form a part of the specification.

Well, there it was. The bare bones of a specification for what would be the world's most advanced antenna.[5] While it was one thing to specify it, it was quite another to build it, and the immediate question was, could it be built, on time and in budget, and by whom?

The AAS was being designed by an iterative process, Stevens told his audience, in the course of which several basic electromechanical configurations had been considered—steerable paraboloid, spherical bowl, various types of smaller antenna arrays, and so on. From these alternative designs, the concept of a single large steerable paraboloid with an azimuth-elevation type of mount had been chosen as most appropriate for continued investigation. Even so, there remained several design options for mountings of the azimuth-elevation (Az-El) type.

In an Az-El mount, the antenna's radio-reflecting surface and its supporting backup framework that serves to maintain its paraboloidal shape are mounted on a very rigid structure known as the alidade. The alidade rotates in the azimuth, or horizontal, direction around a central, precisely known vertical axis. The alidade also carries a pair of elevation bearings that are attached to the backup structure to allow the antenna reflector surface to be driven in the elevation, or vertical, direction. The combination of azimuth position with elevation position allows an Az-El antenna to be pointed in any direction. Basically, it is rather similar to a naval gun mount.

At the time the AAS was being designed, there existed elsewhere in the world two large Az-El antennas that were being used by radio astronomers for scientific research. One had been built in 1957 at Jodrell Bank near Manchester, England, while the other had just been completed in 1961 near Parkes, Australia. Through the good offices of the University of Manchester and the Australian government's Commonwealth Scientific and Industrial Research Organisation, owners respectively of the two radio-astronomy observing instruments, the design features and performance characteristics of both had been made available to the JPL engineers for study and evaluation. The Jodrell Bank and Parkes radio astronomy antennas are pictured here in figures 4.1 and 4.2.

Although it was not then apparent to Stevens (or to anyone else), the final design for the AAS would fall somewhere between these two concepts, but its performance would be remarkably different from either one. On the basis of the engineering studies at JPL, Stevens was confident that a reflector of this size could be designed with sufficient surface accuracy to operate well at the 2.3 GHz frequencies he needed, and that it could be made stiff enough to resist significant distortion as the dish was moved through its full range of elevation. Of more concern to him at the time was the effect of the wind on the servo systems and the huge motors that would be needed to drive the azimuth rotation gear. When erected, the antenna would stand close to 300 feet tall, with a precision reflector almost one acre in surface area. The wind effect would be very important and could be quite different at different heights above the ground.

So they planned to build a 300-foot meteorological tower near the antenna and to equip it with instruments to measure and record the wind spectra—gradients and strength and duration of the gusts. When that was done, they

Figure 4.2. Radio Telescope at Parkes, New South Wales, Australia. This 210-foot-diameter antenna utilizes an Az-El mount on a concrete single-tower pedestal. The azimuth bearings are at the top of the tower, and the elevation bearings are placed at the rear center of the dish. The backup structure for the dish's reflecting surface is clearly visible. The large tripod supports the microwave feed system for the receivers at the prime focus of the paraboloid. The receivers, data handling equipment, control room, and antenna pointing equipment are located within in the tower pedestal. NASA/JPL.

could begin to gather some real wind data to work with. This would lead to a design that minimized wind torque loads and reduced the power that would be required for the huge azimuth and elevation drive systems.

In considering how best to achieve the stated objective for absolute pointing accuracy, Stevens's optical engineers had initially studied the systems used for pointing large modern optical telescopes that had essentially the same problem—accurately following a target moving, at sidereal rate, across their field of view. Since following a moving target accurately was much easier with a small telescope than with a large one, engineers came up with a system whereby the large telescope would be slaved to electronically follow a smaller telescope rather than the target itself. The smaller telescope could be placed in a protected, stable environment and linked via an arrangement of mirrors and light beams and precision servo systems to the larger one. Wherever the small telescope pointed, the larger one would follow. Not quite "smoke and mirrors," but marvelous nonetheless. As it turned out, this idea had been adopted by the designers of the Parkes radio telescope as the preferred method for driving their 210-foot antenna, and they had graciously shared their experience and design particulars with the JPL people. Stevens's engineers were quick to take advantage of their generosity by incorporating these ideas into the absolute pointing system for the Big Dish.

In the center of the azimuth pedestal, they would build a very stable optical tower that would be independent of the antenna structure with all its perturbations and distortions. Shielded by a small enclosure at the top of the tower, a precision optical head would be positioned at the intersection of the azimuth and elevation axes of the antenna. Optical sensors would couple the optical head to a special reference point on the antenna and derive the error signals necessary to cause the azimuth and elevation drive servos to bring the pointing direction of the antenna into coincidence with the pointing direction of the optical head. The optical head, a high-precision telescope like instrument, would receive its pointing instructions from a set of computer-generated predictions of spacecraft position as a function of time. These "predicts" would be based on knowledge of the particular spacecraft's orbit around the Sun as determined by the spacecraft navigation group at JPL. The optical head would eventually be called the "master equatorial instrument," for reasons that we shall come to later. For now, it all seemed straightforward enough. After all, Australian engineers at CSIRO had done it. So JPL engineers could do it.

To this point not much had been said about the radio frequency or microwave system. Stevens had been talking to mechanical engineers, and a major part of the Big Dish design challenge lay in the structural area. But the sole purpose of the antenna was to receive microwave radio signals from distant spacecraft, so the microwave performance was as important as the mechanical performance, and there were significant problems in the microwave area, too. Foremost among them was the question of the exact form, or shape, of the parabolic surface—in technical terms, the ratio of its focal length to its diameter. This would affect the wind loading, the drive power, the weight, the microwave performance, and the location of the microwave feeds for the supersensitive microwave amplifiers or masers (*m*icrowave *a*mplification by *s*timulated *e*mission of *r*adiation) that were a critical element of the overall low-noise receiving system. The masers were fragile and heavy, and required a constant supply of liquid helium and a lot of expert maintenance to keep them running. They had to be mounted at, or near, the focus of the paraboloid where the radio-frequency energy, concentrated by the antenna surface, could be easily fed directly into them for further amplification before passing on to the more conventional receiving equipment. In the Manchester design the preamplifier stages of the receivers were placed on a short tower, and in the Parkes design they were on a tripod, but neither of these antennas used masers, with all the additional complexity that accompanied them. Furthermore, the Big Dish must eventually support a powerful transmitter, and the feed for that, too, would need to be at the focus. There would be a lot of heavy equipment at the focus of the Big Dish, and the focus was a long way from the dish surface. Also, any "stuff" in front of the dish would form an obstruction, generate thermal noise, and generally degrade performance. How was this seemingly intractable problem to be solved?

Again, the JPL designers turned to the technology of optical astronomy for ideas. For many years, astronomical telescopes had used the Cassegrain principle to bring the focus point of the instrument back close to, or even behind, the center of the primary reflector, to provide a more convenient point for making observations. This was accomplished by supporting a small, specially shaped secondary reflector (subreflector) between the center of the primary reflector and its focal point, so that the main beam was, in effect, turned back on itself and could be brought to a focus at any convenient point along the reflector axis at, or near, the center. Such a device is called a Cassegrainian reflector. Stevens's engineers had made use of this principle in the later designs for the 85-foot antennas, and now they planned to extend its application to the Big Dish. A strong, light-weight, conical, enclosed structure mounted on the front of the dish would house the masers and their ancillary equipment and later a transmitter, with the microwave feeds projecting through the truncated top of the cone. By careful design of the subreflector, the main beam would be brought to a focus at that point to deliver the concentrated microwave energy from the reflector surface to the supersensitive masers. The arrangement of Cassegrain subreflector

Figure 4.3. Polar-Mount Antenna at Goldstone, California. This NASA/JPL 85-foot polar-mount receiving antenna shows the Cassegrain subreflector in place, the central feed horn for the maser, and the conical housing for the maser and ancillary equipment. The small, wide-beam antenna mounted in front of the subreflector is used to assist in putting the main, narrow-beam antenna onto its target. NASA/JPL.

and conical housing for the maser and its ancillary equipment is clearly visible in figure 4.3.

Calculations showed that, while there was certainly some degradation in performance from the supporting structure for the subreflector, it was acceptable and was greatly outweighed by the advantages of having a rigid enclosed habitat for the all-important masers.

Stevens concluded, "Because a conservative engineering approach is being pursued, the increased performance of the AAS compared with current machines is an engineering refinement job. No breakthroughs are being pursued to change dark to light. To succeed in this approach we need the following: (1) a careful definition of the technical requirements to avoid asking for something expensive that is not really needed, (2) an accurate definition of the required environmental operating conditions, and (3) the choice of a proper configuration, and great attention to the detail of the design."

In 1961 Bruce H. Rule was chief engineer for the Mount Wilson and Mount Palomar Astronomical Observatories and the Owens Valley Radio Astronomy Observatory in California. He also provided engineering services for the prestigious California Institute of Technology and its Synchrotron Laboratory. Bruce Rule was quite used to working with big, high-precision machines for astronomical applications, and he was Bob Stevens's consultant for the AAS project.

He prefaced his remarks to the engineers assembled at that December 1961 meeting as follows:

> The rapid advancement in the technological fields of communication, radio astronomy, and deep space instrumentation has been so great in the recent few years that the corresponding advanced antenna engineering has hardly kept pace with the new innovations and developments needed for the large antennas. Scientific achievements in space technology, radio astronomy, optical astronomy, and microwave optics have resulted in a cooperative synthesis of many new ideas regarding the required antenna design parameters. Our communication needs have now extended from terrestrial wire or radio links to the near-field telecommunication needs of extraterrestrial projects such as missiles, satellites, and Moon-shot probes. In addition, radio astronomers [are] probing galactic and extragalactic radio "noise" sources, some of which are now known to be at greater distances than can at present be detected even by the 200-inch Palomar-Hale telescope. New knowledge of the universe is coming, to a great extent, from this newest research tool, the radio telescope. (Rechtin et al. 1962)

> In all of these fields of application, ever larger antennas are demanded, having high resolution, approaching coarse optical quality at the higher frequencies, with critical microwave pattern control, having high pointing accuracy, and with precise drive and tracking ability. All of these essential needs must be realizable under the many and adverse ground environmental conditions encountered.
>
> These specific demands are in themselves severe enough for most engineers, but added difficulties include the unknown structural-mechanical limitations of extremely large instruments as well as the ill-defined or misunderstood effects of ground site and environmental factors. One consequence of this is that some knowledgeable designers may be unduly "frightened off" from what they interpret as impossible requirements. At the other extreme is the grossly inexperienced or overenthusiastic designer, who recognizes none of the serious parameters involved, and pitfalls into an untenable design which, at today's costs [1961], can be a serious loss to any institution or user. It is important, therefore, to thoroughly understand the nature of some of the major design problems for large antenna systems and especially for . . . the large steerable paraboloid type.

Understand the problem thoroughly before proceeding. Those were wise words of experience that would be reflected in every step of the JPL approach to the task. It would slow things down somewhat, but it could avoid the costly mistakes to which Bruce Rule had alluded.

He touched on a recent example of this kind of costly misjudgment when he reviewed the principal design features of several of the major radio astronomy antennas then known to be in operation throughout the world. The particular project to which he referred was the 600-foot NRRS radio telescope of the United States Naval Radio Research Laboratory, shown in figure 4.4.

For reasons that remain unclear even today, this project had proceeded through site selection, installation of access roads and facilities, completion of the massive foundations, and erection of some of the azimuth rotating structure, before the design for the entire reflector and its backup structure was fully understood. By the beginning of 1961, construction had progressed to the stage shown in figure 4.5.

By the time that the overall design for the antenna structure was finished, the designers realized that they had a problem. The foundation pedestal, already in place, would not be of sufficient strength to carry the design weight of the antenna and its supporting structure. Reducing the weight, and thus the stiffness or rigidity of the antenna, was not a viable option. Experts from around the

Figure 4.4. Conceptual Drawing of NRRS 600-Foot Radio Telescope. This sketch of the radio telescope to be built in 1961 by the U.S. Navy at its Naval Radio Research Station at Sugar Grove, West Virginia, shows the general arrangement and gives some impression of the enormous size of the antenna and its "Manchester" style of mount. NASA/JPL.

country, including Eberhardt Rechtin, were called in to review the situation and suggest a way to proceed. When it became clear that it would cost about the same amount to redesign and rebuild the base and foundations as it would to incur the penalty associated with canceling the construction contract, the Navy elected to terminate the project entirely. The government was embarrassed over the loss of public monies, and the Congressman for the district including Sugar Grove, West Virginia, was not happy with the loss of jobs and potential development in his area. Whatever the wisdom of this decision, it left the government with a profound aversion to approving funds for the construction of large new antennas, particularly those that went beyond the capability of existing technology.

In retrospect, one might conclude that Stevens's assurances that JPL's Big Dish was essentially only an extension of existing technology, that it was not "changing dark to light," and Rule's insistence on the need to understand the entire project before making the final decision to proceed, were intended to clearly distinguish the JPL approach from the NRRS experience at Sugar Grove. Be that as it may, that was where JPL's Big Dish project stood in December 1961.

Three months later, a United Press International news release datelined Washington, March 8, 1962, reported that NASA had asked Congress for $15 million to build a 210-foot radio antenna capable of receiving "live television from the Moon." The giant dish, to be built at Goldstone, California, would stand about as high as the Capitol Building in Washington, D.C. It was needed, according to Edmond C. Buckley, then director of the Office of Tracking and Data Acquisition at NASA headquarters, "for lunar and planetary missions to be launched in the second half of this decade [1960s]. While the existing equipment [85-foot antennas] was capable of maintaining communications with early unmanned space flights to the Moon, Mars and Venus, the Apollo flights to the Moon in the last half of the century and the unmanned planetary flights of the Voyager program starting around 1966 demanded more sensitive receiving equipment. Both manned and unmanned Moon flights called for the use of television, and television would also monitor the flight of the first Americans to and around the Moon," it said (UPI 1962a). Rechtin's strategy for building a constituency in Washington for a large antenna project had obviously been successful.

Significant though this announcement was to NASA/JPL, it was not the only space-related news competing for public attention at the end of that week in 1962. UPI reported from Cape Canaveral that the United States had achieved its third space success in twenty-four hours by launching a sixth successful Minuteman missile on a 3,000-mile flight to a target in the South Atlantic Ocean. The first combat-ready Minuteman rockets armed with nuclear warheads would be stationed in underground silos near Great Falls, Montana, it said, and eventually

Figure 4.5. Construction on NRRS 600-Foot Telescope, January 1961. The two 420-foot towers were to be used in the construction work and removed when the instrument was complete. The huge box girders between the towers were parts of the eight sets of wheel girders that would support the antenna on two concrete circular tracks. NASA/JPL.

at least 750 such missiles would be installed throughout the western half of the nation. The Minuteman launch had been followed by the launch of an Orbiting Solar Observatory from Cape Canaveral and an Air Force launch of a new "spy-in-the-sky" satellite from Point Arguello in California. UPI also reported the scheduled launch from Cape Canaveral of the "most powerful military rocket ever built in the free world. Standing twice as tall as the 58-foot Minuteman, the mighty Titan II would be able to carry nuclear warheads halfway around the world" (UPI 1962b).

The reports from UPI reflected the peculiar dichotomy of the national interest in the 1960s—namely, the maturing of the Cold War juxtaposed with the dawning of the Space Age. It was in this context that the NASA/JPL Big Dish came to public attention. The first move toward "reaching further out" had been made, and now the race (against time) was on. At JPL there was no alternative but to proceed, and to proceed rapidly, just in case the congressional attitude toward funding for large new antennas changed, as well it might, before the Big Dish could become a reality.

The Contract

Green Bank, West Virginia

The rise of radio astronomy as a new field of scientific endeavor in the decade following World War II was generally attributed to the pioneering research work of three scientists named Karl Jansky, Grote Reber, and John Kraus. In 1933, while working at Bell Labs on a problem related to static noise on short-wave radio communication, Jansky had discovered sources of radio-frequency radiation emanating from discrete regions of the Milky Way galaxy. Jansky was unable to follow up on his discovery, but years later both Reber and Kraus did pursue this intriguing new discovery, each in his own way. In 1937, using a crude home-built radio telescope, Reber made the first systematic survey of radio waves from the sky. After World War II, Kraus founded a radio observatory at Ohio State University and wrote the first standard textbook on the new science of radio astronomy (Ghigo 2004).

However, the substantial cost and resources that were required to build the large antennas needed for research in this new field remained beyond the reach of any single scientific institution. By 1956 the initiative for construction of a national facility for radio astronomy research was taken up by a consortium of research centers known as Associated Universities Incorporated (AUI).

Based in Washington, D.C., AUI was a not-for-profit science management corporation comprised of nine prestigious northeastern universities—Columbia, Cornell, Harvard, Johns Hopkins, MIT, Penn, Princeton, Rochester, and Yale. It had been established in 1946 with a charter to "acquire, plan, construct and operate laboratories and other facilities" that would unify the resources of its members, other research organizations, and the federal government. It was envisioned that AUI would create facilities for scientific research that were too large, complex, and costly to be within the scope of any single institution.

To this end, in 1956 AUI sought to establish a National Radio Astronomy Observatory where it would build the large antennas, the ancillary receiving, recording, and data handling facilities, accommodations, and general logistical support necessary to conduct advanced research in radio astronomy. The choice for the observing site location fell to Green Bank, West Virginia. NRAO headquarters was to be on the grounds of the University of Virginia in Charlottesville (www.gb.nrao.edu).

The tiny hamlet of Green Bank was remote from densely populated areas with their attendant sources of electrical interference. Nearby, a large area of cheap land could be procured and protected from further industrial development, and it was within easy reach of the prestigious scientific institutions where most of the astronomers were established.

NRAO lost no time in establishing a presence in Green Bank. Anxious to take advantage of the excellent radio-quiet observing conditions at Green Bank, scientists quickly set up existing equipment and temporary accommodations and began collecting radio data. The receiving antennas were relatively small dishes, horns, and corner reflectors, but much more ambitious plans rapidly took shape. Two years later Green Bank had an 85-foot-diameter antenna in operation and an even larger 140-foot antenna in the planning stage. The 85-foot antenna was known as the Tatel instrument (figure 5.1), so named for the originator of the conceptual design on which it was based. This, NRAO's first 85-foot antenna, which had been built for NRAO by the Blaw-Knox Company of Pittsburgh, was similar in size and basic design to the NASA antenna then in operation at the Goldstone tracking station in California. Both antennas had been built by Blaw-Knox to the designs of Robert Hall, a brilliant young engineer who at that time headed the Antenna Division at Blaw-Knox.

The November day in 1959 when Robert Briskman and Wallace Ikard drove down from NASA headquarters in Washington to Green Bank turned out to be wet, and very cold. Nature's color printer had not completed the switch from the full-color kaleidoscope of fall to the monochrome palette of winter. Briskman and Ikard were two staff engineers from the Office of Space Flight Operations, a

Figure 5.1. Tatel Radio Telescope, 1999. The first 26-meter (85-foot) parabolic antenna intended specifically for radio astronomy research was built by the Blaw-Knox Company based on a design concept by Howard Tatel and placed in service at the National Radio Astronomy Observatory at Green Bank, West Virginia, in 1960. Still active, the telescope serves as a memorial to Howard Tatel's contribution to radio astronomy. Courtesy NRAO/AUI/NSF.

part of the newly established National Aeronautics and Space Administration. They planned to spend a day or two at NRAO talking with the technical staff there about the possible use of NRAO's larger antennas to track NASA's future planetary spacecraft. For the past year or so JPL had been calling attention to the need for larger antennas for the Deep Space Instrumentation Facility to handle NASA's plans for more ambitious deep space missions. At JPL, Rechtin, Stevens, and Victor had already made technical studies of what might be required and the scale of the costs involved. However, they could proceed no further without NASA headquarters' acceptance of the basic premise that large new antennas were needed, and that NASA should build them. Before that happened, NASA needed to be convinced that there was no alternative to building its own large antennas. Only then would Edmond C. Buckley, assistant director of NASA's Office of Space Flight Operations, take the case to Congress to argue for the necessary funding, and he well knew that Congress would ask some hard questions, first among them being "What are the alternatives?"

It was with all of this in mind that Buckley sent his two engineers down to West Virginia that November to review the NRAO facilities there, and to assess the extent to which NASA might make use of them—assuming, of course, that NRAO would be willing to make its antennas available. This could not be taken for granted. In the closed world of optical astronomy, telescope observing time was a very valuable and carefully husbanded commodity, and in radio astronomy it was no different.

A few days after their return to Washington, they filed a trip report to Buckley (Briskman 1959). They reported that the 85-foot antenna was the largest in operation at NRAO. It had been in full operation, twenty-four hours a day, seven days a week, since being commissioned earlier in the year and was similar in size to NASA's Goldstone antenna. Unlike the Goldstone antenna, however, it did not have the automatic tracking capability to follow an orbiting or planetary spacecraft, nor did it have the microwave feeds and receivers needed for NASA applications. While it would be possible to modify the antenna to suit the NASA requirement, the cost could well exceed a million dollars, and even then the problem of removing and reinstalling the NASA equipment to share the antenna with the radio astronomers appeared to be quite intractable. And there remained the issue of whether NRAO's clientele would be willing to share their precious observing time with NASA in the first place.

There were several other antennas at NRAO at that time. Two of these, a 30-foot and a 12-foot antenna, were too small to be of interest. A third antenna, 140 feet in diameter, was under construction at the time of Briskman's visit. It was expected to be one of the most precise radio telescopes in existence. To obtain such precision, certain mechanical tolerances were very critical. For instance, the dish used a hydrostatic bearing some twenty feet in diameter that had to be machined to a tolerance of five thousandths of an inch in the field. Briskman reported that the contractors were experiencing serious problems in maintaining the critical design tolerances. He planned to revisit the 140-foot antenna a year or so later to see what progress had been made.

While Briskman and Ikard were at NRAO, they also discussed the giant antenna then being built for the U.S. Navy at a site called Sugar Grove about twenty-five miles from Green Bank. Six hundred feet in diameter, this fully steerable antenna was estimated to cost about $70 million and was scheduled for completion in late 1960. Although many aspects of the instrument's capabilities and ultimate purpose were classified, it was known that the Navy required exclusive use of the antenna when the Moon was above the horizon and that access to the site would be severely restricted. Although the capabilities of such an instrument were of great interest, it was obvious that the Navy's stringent access requirements eliminated it from further consideration by NASA. Besides, the antenna existed only on paper. To save time, the Navy had chosen to proceed with the design and construction phases of the project in parallel. While the enormous foundations and facility access roads were well advanced, the antenna itself had not yet reached the final design stage.

Before the NASA engineers left NRAO, there was a short discussion of work known to be in progress at other radio astronomy centers in the United States, Australia, England, and Puerto Rico. There was the 210-foot dish being built at Parkes, New South Wales, for the Australian government. This was known to employ an optical-mirror servo system to position the antenna with great accuracy, and the estimated cost was said to be very modest. There was also Caltech's project to build two 90-foot dishes in California. In England, the University of Manchester had constructed a 250-foot antenna and had been using it for some time to make observations in the one- to several-meter region of the radio wavelength spectrum. Finally, there were several other unique, but as yet untried, designs under consideration in various places. One of these involved a large fixed spherical reflector built into a suitably contoured surface on the ground. The beam was to be steered by a movable radio feed suspended above the antenna by a tower-and-cable arrangement. Such an antenna, a thousand feet in diameter, was already being constructed in Puerto Rico by the Advanced Research Projects Agency (ARPA) through Cornell University under the direction of W. E. Gordon.

The two engineers left Green Bank with the strong impression that, although there were several large antennas in various stages of construction throughout the United States and several in actual operation at Green Bank, none of these were suitable or available for NASA purposes. It was clear that if NASA needed a large antenna for future planetary missions, it would have to build one of its own—there was no alternative. That, they decided as they started through the darkening mountains on the drive back to Washington, would be their recommendation to Ed Buckley (Buckley 1959).

Washington, D.C.

It was early winter in Washington and, although the first snow had not yet fallen, the day was chilly and damp. Black and bare across the city, even the trees appeared to huddle against the cold. Everything, including the sky, was the color of wet concrete.

The recently established National Aeronautics and Space Administration was just entering its second year of operation, so that by the time the visitors found the right address, checked in, got some coffee, and settled around the table, the conference scheduled for that Friday morning started late. Among the attendees were some of the country's leading experts on the topic under discussion that morning, "Large Aperture Antennas for Deep Space Communications." They included William Gordon of Cornell, now responsible for construction of the 1,000-foot antenna in Puerto Rico; Grote Reber from AUI and NRAO, an acknowledged pioneer in radio astronomy research; James Trexler from the Naval Research Laboratory, who was technically responsible for the 600-foot antenna under construction at Sugar Grove; and John Firer from the Carnegie Institute of Terrestrial Magnetism, who brought a wealth of experience in arrayed and reflector-type antennas.

The NASA representatives included Briskman and Ikard, who had visited Green Bank only weeks earlier. Briskman chaired the conference. He explained that after surveying the availability of large antennas throughout the United States, NASA had decided to develop its own specifications for a large antenna that would fully meet its anticipated requirements for support of deep space planetary missions through the end of the century. To benefit from the experience of leading experts in the field, it was seeking advice on what approaches NASA might take in establishing a basic set of requirements for such an antenna. These would be used by one of its field centers for an engineering study that would eventually lead to an industrial contract for the construction of one or more such antennas.

Briskman next explained what functions NASA expected the antenna to perform, and what electrical capabilities it must have in order to perform those functions, not only for the immediate programs but for NASA programs projected far into the future. Then the talk got technical.

They talked about sky coverage, and how it was cheaper if the antenna needed less of it; how bigger was not always better, as far as the "gain" or collecting power of the antenna was concerned, because you might not see any improvement if you could not achieve the tight tolerance required to maintain the precise parabolic shape of the dish. That got harder as the antenna got bigger. Then there was the illumination of the antenna to consider, and that depended on the quality of the radio feed that collected the radio energy reflected from the antenna surface. If the feed could not "see" all the antenna surface evenly, it created side lobes that reduced the "gain" and, worse, picked up spurious thermal noise from the nearby ground surface.

There was also the overarching consideration of what radio frequency or range of frequencies the antenna would be designed for. At that time the DSIF was operating at 960 MHz, but that frequency had not been approved for deep space communications by the International Telecommunication Union at its conference in Geneva earlier in the year. NASA would have to vacate the 960 MHz band and move to the higher frequencies approved by the ITU for space research. What choices of operating frequency were realistic for a large antenna not yet designed and for technology not yet developed? It was beginning to appear that a giant leap of faith would be required here. Discussion of operating frequency naturally led to a discussion of the surface precision required to achieve the desired performance, and ways and means of controlling that precision, as the antenna tipped up and down or was subjected to wind loading—that is, distortion from the pressure of the wind. The higher the frequency, the more precise the surface needed to be and the more difficult it became to avoid distortion. Some of that difficulty depended on the type of mounting that would be used to attach the parabolic surface or reflector, to its foundations. The mounting contained the shafts and drive motors that caused the dish to rotate, to point in the desired direction, and to track a distant spacecraft across the sky from horizon to horizon.

Basically, there were two kinds of mount, each having unique advantages and disadvantages. The azimuth-elevation (Az-El) type was used by the Australian antenna and would be used by the Navy for its Sugar Grove antenna. The hour angle-declination (HA-dec) type had been used on the existing 85-foot antennas at Green Bank and Goldstone with good results. Which mounting would suit NASA's new antenna? The right answer was not clear at this point. Several other approaches to the basic design for large antennas were discussed, but the final consensus seemed to favor the steerable parabolic dish as the approach most likely to result in a successful outcome for NASA.

NASA now had more than enough to think about. Briskman thanked the attendees for their helpful comments and suggestions, and closed the meeting.

Later Ikard summarized the conference for Buckley and identified the recommended several courses of action for NASA. Among the more significant: NASA was to define the actual characteristics required for a large antenna for deep space application and to move expeditiously toward turning the job over to one of the NASA centers with specific recommendations for a method of approach (Ikard 1959).

The response was immediate. Imbued with confidence derived from the Green Bank visit and the supportive expert opinions from the conference on large antennas, Buckley's people began to develop a set of requirements for a large, steerable antenna that would have as much as ten times the receiving power of the existing 85-foot antennas. It would also operate efficiently in the higher frequency bands recently approved for space research by the ITU. JPL would be directed to move its deep space operations into those higher bands as expeditiously as possible. It would employ the most up-to-date microwave technology in the design of its receiving feed system to maximize illumination of the antenna surface and to minimize the susceptibility to extraneous radio noise.

The NASA requirements did not specify a particular design. Rather they called for a phased approach to that end. As they saw it, the task of proposing a feasible design that would meet the performance requirements would be thrown open to all interested and qualified bidders. After the feasibility studies had been completed, a particular proposal would be selected and offered to an industrial contractor for further development into the next stage of detailed design. In the final stage, an open competitive bidding process would be used to select a qualified industrial contractor to carry out the fabrication, erection, and testing of an antenna at the Goldstone site in California.

Fully aware of the accelerating pace of NASA's planetary exploration program, and of the pressure this brought on the overburdened facilities of the existing tracking network, Buckley imparted a sense of urgency to every aspect of the project. He pushed his staff to complete the NASA studies and to document their findings with utmost dispatch. He planned to immediately seek funds from Congress to start actual construction of the antenna in fiscal year 1962. In calendar year terms, that meant starting to spend big money in October 1961. His next step would be to "put the heat" on JPL.

Seen from Buckley's office window on the shady side of NASA's temporary headquarters in the Dolley Madison House, the brilliant green foliage of cherry trees made pools of deep shadow across the sun-bright sidewalks on the opposite side of Lafayette Square. Foot-weary tourists sought the shade for relief from the incandescent afternoon sunshine. August was the peak of the tourist season in Washington, D.C., high summer, hot and humid.

Buckley scanned a two-page letter on the desk before him, shrugged off a lingering shred of doubt, and signed it. It was important that the letter reach Eb Rechtin at JPL in Pasadena as soon as possible. Reminding his secretary to mark it "air mail," Buckley left the office and headed for home.

Eight months had passed since the first NASA conference on large-aperture antennas. During that time NASA and JPL had been fully engaged in reaching agreement on how best to translate their ideas into the reality of a giant antenna in the radio-quiet depths of the Mojave Desert.

Ever mindful of their responsibility to sell the idea to a Congress made wary by the failed Sugar Grove project, the NASA approach tended toward conservatism, reliance on proven designs, and minimal extrapolation from existing technology. JPL in turn, anxious to help NASA persuade Congress to approve the necessary funds, sought to allay their NASA colleagues' concerns with hard analyses to back up their more advanced design ideas.

"It is the purpose of the letter," Buckley began, "to implement the JPL recommendations concerning a proposed 'large' antenna with increased gain, lower noise pickup and higher operational frequency ability than the present 85-foot antennas." Buckley reiterated that technical feasibility studies carried out over the past year at both JPL and NASA had confirmed the need for such an instrument to support the forthcoming NASA program of planetary exploration. The existing NASA antennas were simply not large enough or sensitive enough to receive such weak downlink signals. There was substantial agreement on the concept of a large antenna, and NASA felt sufficient justification to proceed with the project. "Therefore," Buckley continued, "to implement this project we are currently planning to budget $12 million in Fiscal Year 1962 Construction and Equipment funds for the first 'large' antenna" (Buckley 1960).

To the recipient, Rechtin, those were fateful words. The die was cast, the decision was made, the project was on. His fledgling Deep Space Instrumentation Facility would have its large antenna—one for now but, hopefully, three in the years ahead. NASA had done its part. Now it would be up to JPL to build it. Rechtin was anxious to take up the challenge.

Buckley's letter confirmed the procedural agreements that had been thrashed out between NASA and JPL over the preceding months. JPL would initiate a program to provide improved ground antennas for the Deep Space Instrumentation Facility in accordance with NASA's "Requirements for Advanced Lunar and Planetary Programs" document. Initially JPL would take a conventional approach to the design of such an antenna but would concurrently "study promising alternate design approaches." This work was to be carried out jointly by JPL engineers and a suitable industrial contractor. When the study phase had been completed, JPL was to have a "properly selected" industrial contractor produce a conceptual design for an antenna, based on the studies. In the second phase of the project, JPL would "direct a selected industrial contractor in the detailed design, fabrication and erection of the antenna" (Buckley 1960).

Finally, Buckley asked Rechtin to set up a small working group to oversee, coordinate, and plan the project, and exhorted him to make the maximum possible effort in the coming months to "be in a position to start construction of this antenna as soon as the FY 1962 funds become available."

Fiscal Year 1962 would start on July 1, 1961, just under a year away. There was no time to lose.

Merrick

Before the Environmental Protection Agency addressed the problem of air pollution in Los Angeles, the long, hot summer days in Southern California were very smoggy indeed. Jet Propulsion Laboratory, situated in the foothills of the San Gabriel Mountains near Pasadena, was particularly vulnerable. Every summer morning the smog crept up like a noxious brown tide through the valleys from the Los Angeles basin to the foothills, and by midmorning the mountains behind JPL were completely obscured. In late afternoon the smog rolled back out again, leaving the mountains bathed in a radiant glow of pink and orange as the Sun crept behind the western walls of the San Fernando Valley. JPLers generally drove to work early to avoid the unpleasantness of the morning smog.

On such a morning in July 1959, Robertson Stevens left home early for the cross-town drive to JPL with much on his mind. He always enjoyed this relatively short trip in his finely tuned black Corvair. U.S. Interstate Freeway 210 was then still in the planning stage, and the fine, long, straight surface streets of Pasadena provided Bob Stevens with ample opportunity to "exercise the machine," his euphemism for speeding. As he crossed town that particular Monday morning, his mind was not only on the purring Corvair but also on the agenda for the 8 a.m. staff meeting that awaited him at JPL.

The first part of the agenda, status reports from his engineering staff, would be quite straightforward. He expected no surprises there, for he always kept in close touch with what they were doing. But he still had reservations about his decision to name Bill Merrick to manage the design and construction of NASA's "large antenna." Since it would be one of the largest and most precise antennas the world, it would be a very high profile project, and JPL could not afford to fail to bring it into operation on time and on budget. Stevens's own reputation, too, was at stake, for ultimately he would be responsible if Merrick failed. Of the several brilliant young engineers on his staff, he felt Merrick was the best choice. But still Stevens worried. Was Merrick capable? Did he have the right personality? Could he stay the course? Besides that, who would take over Merrick's present job of building more 85-foot antennas if he became project manager for this new Advanced Antenna System?

At the end of Oak Grove Drive, Stevens flashed his pass to the guard on duty and was waved through to the parking lot near his office in Building 125.

In the JPL organization of the time, the Telecommunications Division was responsible for the creation, management, and operation of the Deep Space Instrumentation Facility for NASA.[1] Directed by Eberhardt Rechtin, it comprised several technical sections, each charged with a particular supporting role for the DSIF. Chief of one of these sections, Bob Stevens headed a collection of about fifty engineers engaged in the research and development of the basic building blocks out of which complete systems for deep space communications could be assembled. It was appropriately named the Communications Elements Research Section. Within this section Bill Merrick supervised the Antennas and Optics technical group.

Although it really came as no surprise, Stevens's people applauded the decision to tap Bill Merrick. He came with good credentials and was a popular choice to lead the important new antenna project. He already had several large antennas to his credit and had clearly demonstrated an innate ability to get things done, building a formidable reputation for dealing successfully with the intricacies of government/contractor relationships.

That said a lot about Merrick's personality: he lived with a sense of urgency, he thrived in a crisis when everything had to be done in a hurry, he had a knack for bypassing normal procedures to accomplish tasks that were important by his personal criteria, and he enjoyed the company of top brass. At home he was a devoted family man, interested in all things mechanical, and he was passionately dedicated to the space program. Many years later his daughter, Beth, recalled, "I remember him taking me, as a child, up to the site at JPL to listen to the sound of *Sputnik*. The next day, in an effort to teach me a lesson on my proximity to the making of history, he took me along with several other JPL personnel to dine on hamburgers at a nearby favorite restaurant and celebrate the beginning of the space age. I remember Pickering and von Braun being present. I will never forget what an important day that was, or that he arranged for me to have a small part in it, even if only vicarious" (B. Merrick 2000).

Merrick came to JPL from the Naval Ordnance Test Station at Inyokern, California in 1951. At that time JPL, with Louis Dunn as director, was heavily engaged in testing production versions of the Corporal guided missile system at the White Sands test range as a contractor for the United States Army (Koppes 1982).

Merrick soon became a key member of the White Sands test team that included Eb Rechtin, Walt Victor, and Bob Stevens. His area of cognizance was the optical and electromechanical tracking devices that were used to follow and record the flights of the missiles after launch. It also included the tracking antennas that maintained the ground-to-air radio link for guidance and telemetry signals during flight. The design of accurate pointing and tracking radio antennas became an intrinsic part of Merrick's impressive field of experience and expertise. Within the JPL line organization, he worked under the direction of Bob Stevens as leader of a small group of antenna specialists.

It was natural, therefore, that when the Army called on JPL to build a tracking antenna at Goldstone for its first Pioneer mission to the Moon, planned for December 1958, the task ultimately fell to Bill Merrick. Here was a situation made to measure for Merrick: it came with the approval of authority at the highest level, time was of the essence, it was critical to the space program as he perceived it, and it presented an engineering challenge of unique proportions. That was the motivation for his first visit to Bob Hall at Blaw-Knox late in 1957. That first project made space history, and it also made Merrick's reputation as JPL's expert in the design of large antennas. His second project, the 85-foot azimuth-elevation antenna at Goldstone, enhanced his reputation when Victor, Stevens, and Goldstein used that antenna to bounce radar signals off the surface of Venus and, in the process, refined the known value of the fundamental physical quantity called the Astronomical Unit (Butrica 1996).

Merrick responded to his new appointment with typical energy and attention-getting flair. He created a distinctive name for the group of engineers that would be working under his direction exclusively on the Advanced Antenna System, or AAS, as they called the new "large" antenna within the group. Merrick was a strong believer in the team concept, and henceforth they would be known as the Hard Core Team (see figure 5.2). He believed that giving the team a unique identity that would clearly distinguish its members from their coworkers at JPL would boost their productivity, lead to a more coherent overall design for the antenna, and improve interactive relationships between the many different technologies represented by the team members. It was greatly to Merrick's credit that this concept worked so well, throughout the life of the project, in avoiding most problems and in rapidly correcting those that did occur.

Stevens allowed Merrick a free hand in selecting the members of his Hard Core Team, and Merrick chose the best JPL had to offer. When it was done, the team represented many years of practical experience in all critical areas of engineering and administration. They were called Cognizant Development Engineers. Each CDE was assigned to a particular component of the AAS and was responsible to the project manager for the detailed engineering direction of the design, fabrication, installation, and testing effort in his technical area. The CDEs maintained close liaison with each other to ensure complete integration of the various components as the work progressed at the industrial contractor's plant and at the Goldstone site.[2]

Merrick lost no time in drawing the team's past efforts together. By September 1960, when Buckley instructed Rechtin to proceed, the Hard Core Team had made considerable progress in conceptualizing a basic design for a large antenna and had consolidated these ideas into a coherent design that could be used as the basis for the initial step in Buckley's phased contracting plan.[3]

After reaching agreement with Buckley at NASA headquarters, JPL issued a Request For Proposal (RFP) for a "Giant Antenna Study" in accordance with the current version of EPD 5, and invited a selection of the United States' leading aerospace and high technology construction companies to a two-day preproposal briefing planned for JPL and Goldstone at the end of September. About thirty industry representatives showed up at the briefing. At JPL the attendees were given a general overview of the project, an explanation of the basic performance and constraints required of the antenna, and a briefing on NASA/JPL contracting practices pertinent to the proposal. The next day they traveled to Goldstone to become acquainted with the site (desert), its facilities (nonexistent), and its proximity to habitation (distant). JPL was careful to state the requirements for the antenna in the broadest terms without revealing its own thoughts on the "most probable instrument." The challenge was deliberately left wide open to encourage the broadest possible range of innovative proposals. The contractors were given four weeks to respond.

A Source Evaluation Board met at JPL on November 9 to review the seventeen responses to the RFP and to select candidates for further study contracts (Briskman 1960). In the SEB's opinion, none of the proposals offered any significant advantage in cost, delivery schedule, performance, or risk of success over the team's original ideas for a "most probable instrument," an azimuth-elevation-mounted, Cassegrain-fed, paraboloid-shaped, steerable antenna.

After due consideration of these and other factors relating to contractual matters, relevant experience, resources and industrial history, the board selected four of the seventeen proposals for further study. Not surprisingly, these proposals came from some of the country's top engineering and aerospace firms—Hughes, North American, Westinghouse, and Blaw-Knox—and were seen to offer the most promising design concepts and understanding of the basic problems inherent in the project.

JPL moved quickly to award each of the four companies a three-month contract for a feasibility study. The objective in this first phase was to develop the individual designs to the point where the Hard Core Team could adequately evaluate the technical performance and operational characteristics of the com-

Figure 5.2. Hard Core Team. Members of Merrick's Advanced Antenna System team were photographed in 1962 on the roof of the Space Flight Operations building at JPL. Left to right, they are, *top, standing:* Ireland, McLaughlin, Bartos, Van Keuren (ex officio); *top kneeling:* Valencia, Peterschmidt (ex officio); *bottom, standing:* McGinness, Katow, Stevens, Casperson, Phillips, McClure, Merrick; *bottom, kneeling:* Wallace, Frey, Doster, Lord. NASA/JPL.

peting designs in relation to the requirements identified in EPD 5. To limit the scope of the studies, the team constrained the companies to consider only a pedestal-mounted type of antenna, of limited size and shape.[1] In addition, the designs had to accommodate a data system that would provide not only great precision in pointing the antenna beam but also great accuracy in measuring the actual pointing direction of the beam. Some type of esoteric optical readout technique would be required.

As the various feasibility studies progressed, two significant characteristics of the basic design began to become apparent. First, the "stiffness" of the overall antenna structure—that is to say, its resistance to bending or flexing—would ultimately determine its pointing ability and its efficiency as a collector of radio-frequency energy. This stiffness factor was determined by the alidade structure[5] and by the backup structure, the lattice framework that supported the thin parabolic reflecting surface of the antenna. Second, it would be very difficult to achieve an accurate determination of the direction of the radio beam. The weight of the structural elements themselves was deforming the base on which the angle measuring system was mounted, so that the true position of the radio beam, in space coordinates, could not be measured to the accuracy that was needed for planetary tracking applications.

The Hard Core Team was beginning to wonder whether, collectively, it understood enough about these subtle effects to properly evaluate the efficacy of any given design.

Nevertheless, throughout the three-month feasibility study period, the project appeared to be technically and financially within reach, given good engineering and proper management attention. With Merrick at the helm, the project was assured of that sort of attention. As Merrick later reported, "Both JPL and NASA were satisfied that their intended prime contractual approach seemed to be a good one, and there was no apparent reason why any of the chosen study contractors, or, by the same criterion, any equally competent and experienced contractor, could not proceed with a complete design and fabrication of the prototype instrument" (JPL 1974).

As concern mounted over the structural-stiffness and beam-pointing issues, it became evident that a better understanding of these factors would be needed before the team could produce a comprehensive definition of the overall antenna system that would be adequate for bidding the final construction contract.

In August 1961 the SEB reconvened at JPL to evaluate the four design studies. With recommendations from the Hard Core Team in hand, it selected the Blaw-Knox Company, in collaboration with the Dalmo-Victor Division of Textron Corporation as a major subcontractor, to perform a further feasibility study of approximately one year in duration.

NASA announced the decision in a formal news release on August 25. "Initial feasibility studies," it said, "showed that an increase of three times in the size of the present 85-foot antennas would probably improve communications by a factor of ten, at a cost competitive with alternate solutions involving spacecraft components." Citing Dr. Rechtin, it continued: "Although an improvement on communications by a factor of ten would be useful in deep space exploration at any time, this improvement becomes a virtual requirement with the coming of the Saturn-class launch vehicles. These vehicles are capable of launching spacecraft that can perform missions that were not possible before. Typical of these new missions are lunar roving vehicles, photographic lunar and planetary probes and advanced programs associated with manned flight to the Moon" (NASA 1961).

This phase of the Blaw-Knox feasibility and design study was expected to be completed by July of 1962, and the planning schedule called for the 240-foot antenna to be operational at Goldstone by January 1965. The amount of the one-year contract was to be $250,000 (NASA 1961).

Displaying his remarkable prescience in this public statement, Rechtin had associated the new antenna more with the manned lunar program than with furtherance of planetary exploration. This, he hoped, in light of the gathering interest in the manned space-flight program, would increase its appeal to Congress when the time came to approve the funding for the antenna.

The terms of their new one-year contract instructed the Blaw-Knox/Dalmo-Victor team to conduct a feasibility study aimed at finding the best specific configuration for the overall antenna system. Such matters as transportation from the subcontractor's plant and assembly and erection of the complete antenna at the remote Goldstone site were to be included. A complex mathematical error analysis, based on reasonable error budgets for each component, was required to show that the overall system met the performance specification set for it.

As the study progressed, the Hard Core Team became more confident that the azimuth-elevation type of design offered considerable advantages over all other types of mount, because the antenna reflector would be subjected to gravity loading about only one axis (elevation), rather than about multiple axes as in all other types of mount that were considered.

In the Az-El design the azimuth rotating parts of the antenna (alidade) were to be supported by a circular pedestal with a ring girder to simplify the precision azimuth-angle readout system and to minimize deflections as the antenna rotated in azimuth. Earlier it had been determined that a large hydrostatic thrust bearing offered the best arrangement to transfer the weight of the entire moving structure to the pedestal. To carry the wind-shear loads, a radial bearing was preferred over the roller-bearings-and-king-post arrangement that was commonly used on smaller antennas.

Fred McLaughlin assured the team that the designs for gears and reducers[6] presented little problem in achieving the desired stiffness. They were to be used in pairs to minimize the effects of backlash, and the size of the gears and bull gears could be optimized for cost and other parameters.

Since it was anticipated that the azimuth rotation system would contribute most to the lack of stiffness of the overall system, this was subjected to particularly stringent analysis. For that purpose, Smoot Katow developed a complex analytical model of the azimuth rotation system using one of the earliest digital computers to arrive at JPL. Sure enough, Katow's analysis showed the alidade structure would be the limiting element in the azimuth rotation system because of its low stiffness. Although the hydraulic motors, servos, and gear train that comprised the azimuth drive system could easily attain the degree of stiffness required by the specifications, the alidade structure would have to be made extremely rigid to avoid degrading the overall performance of the azimuth rotation system.

Houston McGinness, who directed the analysis of the optical angle-data systems, put forward two critical stipulations. First, the two axes about which the antenna reflector, or dish, would rotate—azimuth and elevation—must intersect. Second, there must be a clear, uninterrupted line of sight from that point of intersection to a ground-based reference point. This arrangement not only simplified the placement of the optical elements that were used to determine the pointing direction of the antenna beam, it also tied the dish position to a fixed reference point on the Earth's surface via an optical path and an inertialess beam of light. Translated into hardware, this required a very rigid instrument tower that would be concentric with the pedestal and tall enough to support the optical elements of the angle-data system at the intersection of the axes of rotation. The reference point would be located internally, at the center of the tower base. Complex though it seemed, it was less complex than other configurations where the axes of rotation did not intersect. And, perhaps more important, it was based on a similar system that had been developed successfully for the new 64-meter (210-foot) radio astronomy antenna then being built near the rural town of Parkes in Australia (Stevens 1961).

As the various designs developed, Stevens furnished NASA headquarters with regular progress reports, and by the end of the year when Merrick brought the Hard Core Team together to make the final decision, it had become quite obvious which of the overall designs came closest to meeting the criteria established for each of the systems. Merrick took the team's recommendation to Rechtin, Victor, and Stevens. All three agreed that the best approach to an overall design had at last been found. Rechtin quickly dealt with the formalities of bringing NASA headquarters into the decision-making process, and approval to proceed to the final phase, initiation of a contract for the construction of a large antenna at Goldstone, followed shortly after. JPL then instructed the Blaw-Knox/Dalmo-Victor team to prepare the essential drawings and documentation to define the final configuration, and the formalities of issuing a Request For Proposals for the actual construction contract began. It was January 1962.

The UPI newswires for March 8, 1962, carried two items of particular interest to the United States space program, one related to the military, the other to NASA. The items vied with one another for lead story space in the local print media. In Los Angeles the *Times* took the military article while the *Herald-Examiner* chose the NASA item.

Datelined Washington, March 8, its story said, "The Space Agency is asking Congress for $15 million to build a 210-foot radio antenna capable of receiving 'live television from the Moon.' The giant dish, to be built at Goldstone, California, would stand about as high as the Capitol building. It is needed, according to NASA, for lunar and planetary space missions to be launched in the second half of this decade. Edmond C. Buckley, director of tracking and data acquisition for NASA, told Congress today that existing equipment is capable of maintaining communications with early unmanned space flights to the Moon, Mars and Venus. The largest antennas now in use are 85-footers. But the Apollo manned flights to the moon in the last half of the century, and the unmanned planetary flights of the Voyager program starting around 1966, demand more sensitive receiving equipment." The story concluded, "Both manned and unmanned flights call for use of television. Television will also monitor the flight of the first Americans to and around the Moon" (UPI 1962a).

Despite the fact that there yet was no final design for the large antenna and a number of very significant problems remained to be solved, Buckley's statement reflected NASA's confidence in the eventual outcome. He had indeed "put the heat" on JPL.

Merrick turned over the task of implementing the two remaining 85-foot antennas in the DSIF, one at Woomera, Australia, the other near Johannesburg, South Africa, to others in his group in order to devote his full attention to the Advanced Antenna System. Driven relentlessly by Merrick, the Hard Core Team worked frantically to resolve the remaining issues and revise EPD 5 to reflect the new ideas, while the Dalmo-Victor engineers struggled to complete the drawings and documentation needed for the pre-procurement briefings for the final construction contract now scheduled for mid-May at JPL.

As a measure of the interest this project had developed in the aerospace industry, thirty-four companies sent representatives to the two-day briefings and conferences at JPL and at Goldstone. JPL followed the same agenda it had used for the feasibility studies contract. JPL contract administrators explained the applicable NASA procurement and contract formalities, while Stevens and Merrick, supported in some areas by specialists from the Hard Core Team, addressed the technical, engineering, and logistic aspects of the project. Then the attendees traveled to Goldstone to inspect the actual site for the antenna before returning to their respective companies to prepare their bids for a fixed-price contract for the fabrication, erection, and testing of the large antenna as specified in the second revision of EPD 5.

Despite the initial show of interest in the project, only three companies actually responded with proposals, and none of them came within negotiable range of the funds that were available for construction of the basic antenna structure and mechanical system. In fact, the average bid was 40 percent greater than the funds available. Disappointed though it was with this result, the Hard Core Team immediately began an intensive review of the project requirements as they were then stated in EPD 5. Representatives of the antenna construction industry were called in to advise on practicable ways and means of reducing the overall cost. After much debate, the engineers agreed on a way to modify the specifications without unduly compromising the original requirements for technical and operational performance. Once again EPD 5 was revised, and a new request for proposals was issued (Merrick, Stevens, and Rechtin 1963).

Nine firms expressed an interest in the redefined project. Four of the nine companies responded with proposals: North American Aviation, McKiernan-Terry Corporation, Rohr Corporation, and the Blaw-Knox/Dalmo-Victor companies. This time, the SEB found all four proposals responsive to the basic intent of the restated requirements and within negotiable range of the funds available for the project.

Late on a Friday afternoon in January 1963, the peaks of the San Gabriel Mountains behind JPL still retained their dusting of light snow from the previous night when Merrick called an all-hands meeting of the Hard Core Team. In Merrick's terms, an all-hands summons meant "drop everything, come to the Building 125 conference room, something of great importance to everyone involved with the AAS is about to be announced." This was no exception. With his usual tight-lipped grin rather more expansive than usual, Merrick passed the word to his team that the NASA administrator had instructed JPL to negotiate a contract for the construction of the advanced antenna with the Rohr Corporation.

The official NASA news release soon followed. "NASA announced today the selection of the Rohr Corporation of Chula Vista, California, to design, manufacture, install and test an advanced antenna system to be located at Goldstone, California. The contract amount for this work is about $12 million." The announcement went on to say that the antenna, 210 feet in diameter, would be similar in appearance to the Australian (radio astronomy) antenna at Parkes, New South Wales. It would, however, have major improvements that would permit operation in adverse wind conditions. The new antenna represented the culmination of three years' study and design work, and was expected to improve the communications capability of the Deep Space Network[7] by a factor of ten over the existing network. NASA named JPL to manage the contract, with completion expected in thirty-six months (NASA 1963).

NASA's announcement of its plan to build a huge new antenna at Goldstone was carefully worded to avoid any appearance of upstaging the Australian designers of the Parkes antenna. The Australians had been very cooperative with JPL throughout the design process, and even permitted Merrick to conduct some engineering tests on their structure and servo systems. The JPL design adopted several of the Parkes system designs, most notably the optical angle-data technology. The JPL design had also benefited from the Australian experience with ideas that had not worked out for them, or had been rejected in their own design process. In addition, the use of the Parkes antenna to augment NASA's own facilities in Australia was a distinct possibility for the future. These ideas must have been in Rechtin's mind when he elected to restrict the diameter of the AAS to 210 feet (64 meters), the same size as the Parkes antenna, despite the protests of Merrick and Stevens that they could design it to be somewhat larger. It was deliberately described as "one of the largest precision antennas in the world." As NASA downplayed its rugged, monolithic design by noting that it was "similar in appearance to the Parkes antenna," the reference to its superior wind-loading performance could not offend.

Finally, in emphasizing its benefit to the planetary program and omitting reference to support for the Apollo program, the announcement reflected the changed outlook at NASA with respect to JPL's prime responsibility—managing NASA's unmanned lunar and planetary program. The announcement was also timely in another way. Although the Mariner flyby of Venus in December had been a great success, the loss of five successive Ranger lunar missions earlier in 1962 still soured JPL-NASA relations. The announcement of a bold new venture on Earth went some way toward reasserting NASA's confidence in JPL and ameliorating the strained relations.

There is no record of the reactions of the unsuccessful bidders in the aftermath of NASA's decision. Undoubtedly the Blaw-Knox/Dalmo-Victor team must have believed that, in view of its work in developing the final designs for the 210-foot antenna and its long association with JPL in providing the 85-foot antennas for the network, it had the best chance of securing the contract. It must have questioned NASA's choice of Rohr, when relevant experience was considered and both bids were within negotiable range of the funds available. Whatever the basis for NASA's decision, it remained intact, and a fixed-price contract for the design, construction, and testing of one of the "largest precision antennas" was executed with the Rohr Corporation on June 20, 1963.

In some minds the questions persisted: What was the definitive attribute that led NASA to choose Rohr over Blaw-Knox for that prestigious contract? Hindsight suggests that the answer lay not so much in Rohr's corporate experience in the antenna industry as in the wisdom of an intuitive personnel decision made two years earlier by its CEO Frederick Rohr.

Rohr

In May 1927 the name of Charles A. Lindbergh stood tall among the pioneers of early aviation. As the first person to fly an aircraft directly across the Atlantic from New York to Paris, he won a dominant place in aviation history and created a huge surge of interest worldwide in the new business of aviation.

Much was subsequently written about the daring pilot and his durable aircraft, a Ryan model NYP built by the Ryan Airlines Corporation of San Diego. The plane was a modification of a standard design, done at Lindbergh's direction to accommodate a specially built large-capacity fuel tank. The innovative tank was, of course, a critical factor in making such a long flight possible. It was built at the Ryan plant under the direction of a young engineer by the name of Fred Rohr.

As the aircraft manufacturing industry in San Diego began to expand with the onset of World War II, Fred Rohr founded his own company, Rohr Aircraft Corporation, and introduced the system package concept for aircraft production, specifically the power-plant systems, which included engines, electrical and electromechanical components, and engine cowlings. He developed advanced methods of drop-hammer forging, use of stainless steel honeycomb structures, and a new technique of stretch-forming sheet aluminum to form the complex curved surfaces of aircraft engine cowlings. Over the next twenty-five years, under its CEO Frederick H. Rohr, the Rohr Aircraft Corporation became a very high-class contractor for the aircraft-aerospace industry.

In 1960 "Pappy" Rohr, as he was affectionately known by then, was looking for ways to expand his company's existing interest in small antenna construction when the NASA/JPL intent to seek a contractor for a large new antenna came—or, more accurately, was brought—to his attention. Here was a project that would require the manufacture of a large number of precision-shaped aluminum panels with very exact specifications in addition to a substantial number of electrical and electromechanical system packages, all of which was well within the range of Rohr's expertise. The company already had in place much of the specialized fabricating infrastructure and subcontracting control processes that such a large project would require. What it did not have was a manager with the experience necessary to lead a new project of that size. But with an intuition based on years of experience in the aircraft industry, Pappy Rohr knew where to find one. He vowed to take action on this matter before the year was out. Then one day in late November 1960, before he actually did anything about it, he received a phone call from Pittsburgh, Pennsylvania. It was a young engineer, currently with the Blaw-Knox Company, who had a great reputation and a brilliant idea. His name was Robert Hall.

As the United States returned to peacetime pursuits in the post–World War II years, television replaced radio as the dominant medium for home entertainment. To meet the growing popular demand for service, TV stations needed more powerful transmitters and towers of ever increasing height to maximize their service areas. Out of this latter demand, there rapidly developed a sizeable industry specializing in the design and manufacture of tall masts and towers for television and radio applications. It was Blaw-Knox engineers that built the first thousand-foot-high television tower in the United States, and it was Blaw-Knox that built the microwave towers for the AT&T transcontinental communications system between New York and Los Angeles in the early 1950s. The Blaw-Knox Company of Pittsburgh, Pennsylvania, soon emerged as one of leaders in that industry.

A key figure in the early success of these projects was a recently arrived engineer in the radio and TV tower department at Blaw-Knox by the name of Robert D. Hall. Later, Hall was given charge of a new Antenna Structures Department, where he helped Blaw-Knox win a contract for the design and construction of 316 parabolic antennas and mounts for a Tropospheric Over-the-Horizon radar system on the DEW line—the U.S. military's chain of Defense Early Warning stations on the Arctic Circle. By the late 1950s he had become one of the country's leading authorities on large antenna structures.

Early in 1957, two scientists from the Department of Terrestrial Magnetism at the Carnegie Institution of Washington approached Robert Hall with some new ideas for a large antenna to search for the cosmic radiation in the central regions of the Milky Way. In effect, what Merle Tuve and Howard Tatel wanted was a steerable, polar-mounted antenna that would work efficiently at a frequency of about 1824 MHz. That was no small order in 1957. Hall said he would think about it.

In a few months he had turned the ideas of Tatel and Tuve into a practical engineering design for a steerable, polar-mounted, 85-foot-diameter antenna with a parabolic surface contour that was accurate to within one-eighth of an inch, quite satisfactory for collecting radio energy at a frequency of 1824 MHz. Once the manufacturing design was complete, Blaw-Knox went looking for buyers. The first buyer was the National Radio Astronomy Observatory at Green Bank, West Virginia. Blaw-Knox built and installed their antenna in 1958. It became known as the Tatel antenna after the originator of its conceptual design.

Word of the new polar-mount radio astronomy antenna being built for NRAO spread rapidly through the scientific community and it was not long before the University of Michigan called on Blaw-Knox for a similar antenna. By December 1957, fabrication of the component parts for the second 85-foot antenna was well under way at Blaw-Knox when Bob Hall received a visitor from the West Coast. He said that he represented the Jet Propulsion Laboratory of the California Institute of Technology and his name was Bill Merrick. Although Hall could not have known it at the time, the course of his engineering career had just taken a sharp change in direction (R. D. Hall 2000).

Merrick was looking for a large steerable antenna that would be a critical part of the United States' initiative to launch a space probe to the Moon in December 1958, as a powerful answer to the Russian *Sputnik* spectacular of the previous October. He wanted it now, and he asserted that he had connections in high places that would insure that he got what he wanted. His first question—"Can you modify this antenna design to track the radio transmitter of a space probe at 2295 MHz?" (instead of natural sources of cosmic radiation at 1824 MHz)—set Bob Hall thinking. He knew that careful attention to the design of the surface contour would allow the antenna to work fine at the higher frequency, but he needed to understand the mechanical problems related to the much greater demands for accurate antenna pointing for space probes. He called his junior engineer and best draftsman, Fred McLaughlin. "Take Merrick's ideas on beefing up the antenna structure, add bigger drive gear-boxes, and draw up a modification proposal," he ordered (McLaughlin 1999).

While McLaughlin was busy with the modifications, Merrick raised the question of delivery time. He needed it, he told Hall, to be completely erected on-site at Goldstone, aligned, and tested in time to meet the scheduled December 3, 1958, launch date for the Pioneer lunar probe. That left ten months for the entire project.

"The only way that we could do that," said Hall, "would be to take components already fabricated for the Michigan antenna and use them for the JPL antenna. But that would make the Michigan astronomers very unhappy."

"I'll take care of that," replied Merrick as he took McLaughlin's drawings and returned to JPL. The JPL contract, which included the Michigan components, arrived at Blaw-Knox a short time later.

In due course the 85-foot Blaw-Knox antenna was erected at Goldstone on what became known as the Pioneer site, just in time to track the first of the U.S. Army's Pioneer lunar probes.

Although Blaw-Knox was the prime contractor, it relied on a number of subcontractors for design and fabrication support in all of the specialized areas in which it lacked the necessary expertise. Chief among these was Dalmo-Victor, a company from the San Francisco area that specialized in servomechanisms and precision control systems. Blaw-Knox/Dalmo-Victor soon became a strong working partnership in designing and building big antennas, structures, and drive systems. Under the engineering direction of Bob Hall, they had done good work on the Goldstone antennas, and at JPL both he and they were held in high regard by Merrick and upper-level managers of his department.

It was through this association that JPL's plans for another, much larger, antenna came to Hall's attention, and he expressed a strong interest in bidding on behalf of Blaw-Knox for a contract. As a result Blaw-Knox, along with sixteen other major industrial contractors, responded to JPL's initial request for a conceptual proposal for a large steerable antenna, approximately 200 feet in diameter, that would be capable of tracking space probes out to the farthest reaches of the solar system.

While Merrick and his engineers at JPL were reviewing the various submissions of conceptual designs, Bob Hall made a critical decision regarding his personal involvement in the antenna construction industry. As he saw it, his future lay in designing even larger spacecraft-tracking antennas, with improved beam-pointing capabilities. Such antennas would have to be much stiffer, so they would not distort from their own weight as they moved in the elevation direction, and they would need very precisely shaped parabolic reflecting surfaces to produce the incredibly narrow radio beams required to track distant space probes. He decided that, rather than being in a steel-plant-related industry like Blaw-Knox, where there was little optical tooling or anything like that to make accurate surface measurements, he needed to be with a company that was directly involved in the manufacture of accurate curved-surface aluminum structures, and that had good basic design and manufacturing skills in metal forming, servo systems, hydraulics, and electrical control wiring.

He buzzed his secretary. "Get me the Rohr Aircraft Corporation in San Diego, California. I want to speak with the CEO. His name is Frederick Rohr." Hall had a proposition for him. The timing could not have been more propitious (Hall 2000).

He and Pappy Rohr talked for an hour. When the phone call ended, Hall buzzed his secretary again. "Get me on tonight's red-eye flight to San Diego." A week later he returned to Pittsburgh with an agreement from Rohr that would expand the corporation's existing antenna department into a large Antenna Division for the specific purpose of designing and building large-diameter, high-precision radio antennas for space probe tracking purposes. He would be its engineering manager, and he would take up the position as soon as he could arrange the move to California. Hall left Blaw-Knox in December 1960, took a month to settle his affairs, and moved to Chula Vista near San Diego early in the new year. Just two years later, the Rohr Corporation was awarded the prime contract for NASA's first 210-foot antenna (R. D. Hall 2000).

Did the reputation of Robert Hall influence the decision? Might the decision have been different had Hall stayed with Blaw-Knox? Did Hall's association with Merrick and others at JPL affect the outcome? Was the Rohr Corporation's location in California a factor? Or was it, as Hall had presciently observed, that precision sheet-metal fabricating technology and well-established optical facilities for curved-surface metrology, such as he had found at Rohr, were perceived to be key factors in the construction of large precision antennas?

Whatever the answer to those questions, Pappy Rohr's decision to bring Hall to Rohr was well and truly vindicated. But Pappy knew only too well that winning a contract was one thing, completing it was quite another. Hall remembered meeting Pappy Rohr one day early in 1963, right after the contract was signed. "I hope we can make good on that contract," Pappy remarked. "Don't worry, Pappy," Hall replied. "We'll make good on the contract, and make Rohr some money, too." As we shall see in the course of this narrative, he did both (R. D. Hall 2000).

6 Goldstone, California

The Big Hole

Well before dawn on a December morning in 1963, three engineers huddled in their winter jackets against the biting cold air of the Mojave High Desert and stared anxiously at a small pile of wet concrete as it slumped slowly below the reference mark on the test fixture. To one side of them, brilliantly floodlit, an intricate web of high-strength steel reinforcing bar framed a gaping circular hole in the desert floor. Behind them an enormous concrete-mixer truck waited, its diesel idling, for their approval to discharge its load of six cubic yards of high test concrete to the bottom of the hole below. This was the first of hundreds of cubic yards of special concrete that would be placed in the hole that day to form a continuous concrete ring, 100 feet in diameter, 11 feet wide, and 3 feet deep. When completed, it would form the foundation for the pedestal of Goldstone's new 210-foot-diameter Advanced Antenna.

The concrete for the foundation and pedestal was mixed in a carefully constructed "batch plant" a short distance from the antenna site. Cement, sand, and specially selected aggregate from a quarry near Barstow had been trucked out to the desert site for mixing, in precisely controlled proportions, at the batch plant. Water of low mineral content was also brought twenty miles to the site from Fort Irwin. A fleet of five mixer-trucks shuttled the mixes from the plant to the site in a continuous stream to ensure that the entire foundation was completed in a single, uninterrupted pouring operation. This was no ordinary concrete. To carry the enormous load of the antenna and its ancillary drive systems without deterioration or deformation, it needed to be of the very highest quality. Particular attention had been paid to selecting a high-strength aggregate, the crushed rock that gave strength to the cured concrete, and to the temperature of the mix as it was placed in its formwork. This being winter, Don McClure had set up a propane heating system under the water storage tanks to warm the water to the desired temperature before mixing with the cement and aggregate.

That winter morning the first batch to pass the "slump test" was placed before sunrise. The last batch of the single "pour" went in fourteen hours later. It was a long day.

After the foundation was laid, many such days were required to complete the pedestal and instrument tower to the exacting standards set for the quality of the concrete. By October 1964 the job was done (figure 6.1), and the foundation, pedestal, and instrument tower were ready to carry the loads for which they had been designed. It was time for "big steel."

Ron Casperson was the Hard Core Team expert on steel, big steel. He had been involved in the project from the very beginning, as early as December 1960. Merrick was anxious to have a licensed civil engineer on his staff. Casperson had accumulated years of experience with Kaiser Steel in California and with Peter Kiewit working on the big Garrison Dam in North Dakota, and he was also a licensed civil engineer with a general engineering contractor's license from the State of California. It was a perfect match, not only in qualifications but in temperament. Casperson adjusted rapidly to Merrick's unusual, and at times exasperating, style of management. He became Merrick's first lieutenant throughout the entire project. He regarded his boss's quirks quite objectively and often soothed the hurt feelings of engineers who were upset by Merrick's unintentionally irritating actions. He had an endless repertoire of Merrick-related stories, not all of which bear repeating. One of his favorites recalled an occasion during the early construction of the first 85-foot antenna at Goldstone when Merrick rolled a lighted cherry bomb under the door of the washroom while one of his engineers was sitting on the toilet. Casperson was called upon to convince the startled engineer that the firework was a prank, not an assassination attempt. Among his colleagues, Merrick was well known for his practical jokes (Casperson 2000).

If Ron Casperson was Merrick's first lieutenant, Don McClure was Merrick's first mate. As JPL's on-site manager, he resided at Goldstone and represented Merrick there. Nothing happened at the Goldstone job site without McClure's knowledge and approval. He monitored the progress of the work being carried

Figure 6.1. Construction at Mars Site, October 6, 1964. The antenna pedestal structure has been completed and work continues on the interior working spaces. The concrete portion of the central instrument tower is complete and ready for installation of the steel upper section. The base of the 300-foot-high tower derrick is visible to the left of the pedestal. To the right are temporary work trailers and permanent buildings for the diesel power generator, pump house, and water cooling tower. NASA/JPL.

out by all of the contractors and worked closely and amicably with his counterpart from the Rohr Corporation, Don Ramsey, to resolve problems and settle disputes. The full resources of the Hard Core Team, including Merrick himself, were at his disposal at all times to resolve matters on which he required assistance. He reported progress daily to Merrick back at JPL and took whatever action was needed at Goldstone to ensure that the planned construction schedule was being met. He recalled that when he took the job, Merrick said to him, "Your job is not to solve design problems; if one comes up, your job is to know who to call." That is what he did, but he later admitted that "there was a lot of on-the-job learning" (McClure 1999a).

McClure joined JPL in 1961 to supervise the technicians from Merrick's Antenna Structures and Optics Group who were working on the early 26-meter antennas then under construction at Goldstone. About a year later, when Merrick sought a resident engineer for the AAS project at Goldstone, McClure's general engineering degree from UCLA, familiarity with JPL antenna construction and testing, and availability in the Goldstone area (he resided in Barstow) made him a natural choice.

In retrospect, he thought Merrick was "a brilliant engineer, but he had poor people skills. He was very smart, very task oriented, and to him, the project was it, the people were secondary, the job was his whole life. Nevertheless," said McClure, "around JPL he was called Wild Bill Merrick for a reason. He could be real wild about the things he wanted done, or the way he wanted them done, but he had brilliant ideas—not always, but enough so that people respected him for it. He was well regarded by the team" (McClure 1999a).

McClure's first task was to extend the existing paved road five miles across the desert to the new construction site. Before such work could start, however, McClure built a large plastic-lined pond near a disused well to hold the water that the road builders would need for the roadwork. It held about a million gallons of rather saline water and immediately became known at Goldstone as Lake McClure. The much purer water required for the pedestal concrete was trucked in over the new road from Fort Irwin, about twenty miles away.

With the access road complete, work began on the huge excavation required for the antenna pedestal and the instrument tower. Figure 6.2 gives an idea of the size of the pedestal and instrument tower in relation to the overall antenna structure.

McClure was a cheerful man, always ready to hear an opposing opinion in an argument before making a decision on how to proceed, and he was endowed with a wry sense of humor. When the excavations were finished, McClure had some aerial pictures taken of the AAS site. He titled one of them *AAS Hole 1964* and had it framed for hanging in his Goldstone office trailer.

The completed pedestal consisted of a reinforced concrete cylinder about 80 feet in diameter with a wall thickness of 3 feet 6 inches and a flat slab top having a circular collar in the center. It weighed 5,000 tons. The collar surrounded a circular opening through which the instrument tower would pass. The pedestal provided support for the moving section of the antenna and contained ample interior space for work areas and facilities. The instrument tower consisted of a completely independent concrete cylinder with a silo-type foundation extending to 35 feet below grade level. The concrete extended up to the top of the pedestal. A steel cylinder was added later to bring the tower up to the level of the elevation axis. It weighed 600 tons and exactly matched the weight of the soil that was excavated to support it. This ensured that the tower, which supported

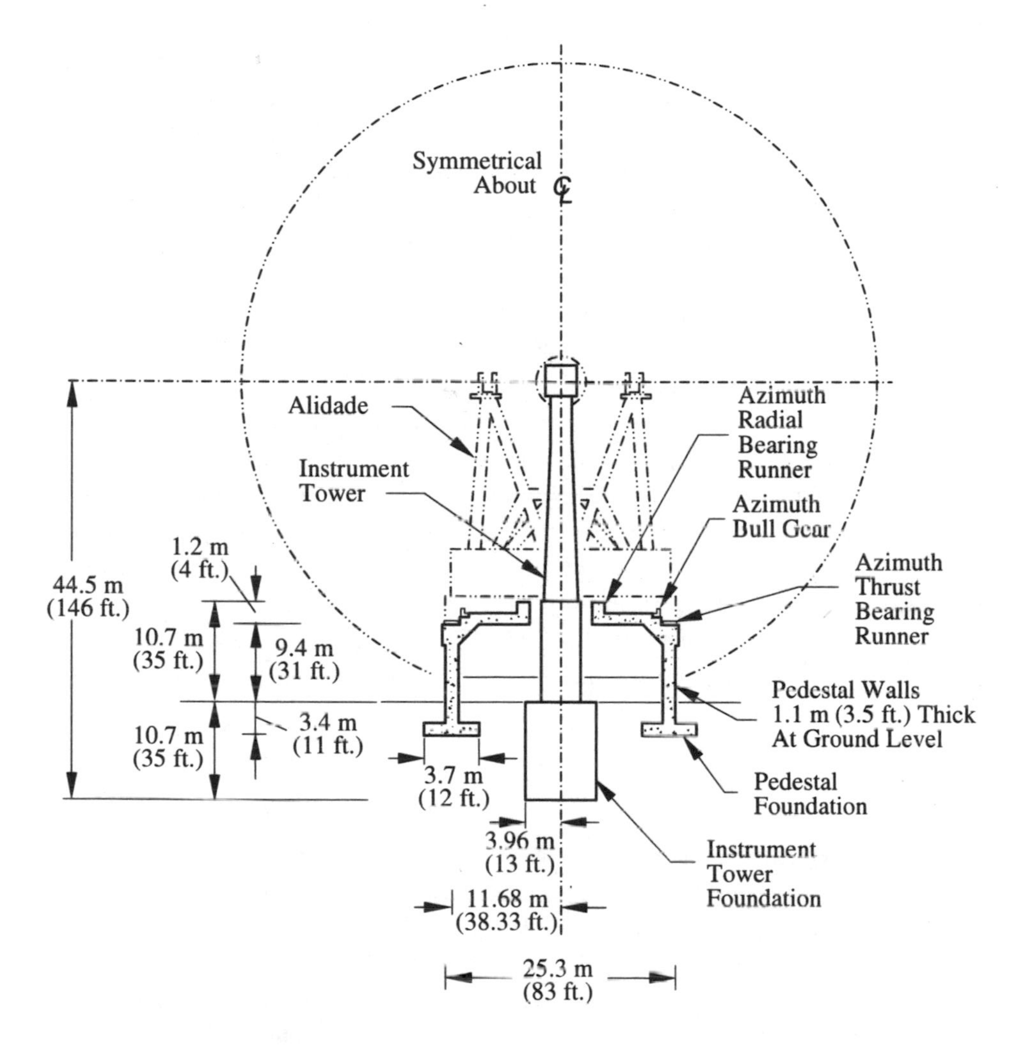

Figure 6.2. Cross-sectional Diagram of Pedestal and Instrument Tower in Relation to Overall Antenna Structure. The instrument tower required a hole 30 feet in diameter and 40 feet deep. The pedestal hole, concentric with the instrument tower, was 11 feet below grade and 100 feet in diameter. Illustration by Fred D. McLaughlin, NASA/JPL.

the precision angle data system, would have the maximum possible stability in its foundation.

McClure remembered well the huge amount of concrete and the problems he had with it. "The concrete used in the pedestal was designed to be very stiff, to match the stiffness of the steel runner of the hydrostatic bearing. So we had concrete with a very low water content, and every batch was tested to verify it was right before it went into the hole. When it cured, that concrete was like granite. I had a piece polished for an ornament."

Despite the painstaking care that was taken with its mixing and placing, however, the composition of the granitelike concrete was to have a near disastrous effect on the capability of the antenna a few years later. Unknown to anyone at the time, the concrete contained a "time bomb."

While the squat cylindrical shape of the concrete pedestal slowly emerged from the desert floor, a lot of other things began to take shape nearby. Most prominent on the site was the tower-derrick crane that Rohr planned to use for assembling the heavy steel components of the antenna structure. With a height of 315 feet and a lifting capacity of 200 tons, the derrick was carefully sited so that it could place all the components of the antenna in position as construction progressed. Tall towers were Bob Hall's specialty, and this was a very tall tower. Don McClure remembered the enormous concrete anchor blocks, 8 feet on a side, that he buried 10 feet underground to secure the guy wires that held the tower in a vertical position.

Gradually, permanent buildings began to appear on the site. They would house three 500-kilowatt diesel-engine electric generators to power the new station, high-pressure pumps for the hydraulics, and water chillers for the electronic equipment. Off to one side, a storage area held steel components of various sizes ranging from large to very large, waiting their turn for assembly on the main antenna structure. Most prominent among the big pieces of steel were eleven slightly curved segments of the hydrostatic thrust bearing. Each 11-ton segment consisted of a finely machined slab of steel 22 feet long, 44 inches wide, and 5 inches thick. When assembled end to end on top of the pedestal, these segments would form a circular base or runner 80 feet in diameter and perfectly flat. The runner, looking very much like a huge steel washer, provided the base around which three hydrostatic bearing pads carrying the full weight of the antenna structure, all 5 million pounds of it, would move with great precision to impart azimuth motion to the antenna.

The principle of hydrostatic bearings was not new—at that time they were in fairly common use for high-precision machinery applications—but this was the largest hydrostatic bearing that had ever been designed and built. How well it would work in this extremely demanding application remained to be seen. Everyone concerned with its development realized that it was "pushing the envelope," and watched its assembly and alignment with some apprehension.

Basically, the hydrostatic bearing contained four simple parts—a runner and three pads. The runner consisted of a very flat circular steel surface mounted securely to a heavy concrete base, while the pads were rectangular steel blocks that could move freely around the runner. The three pads, located at the corners of a triangular frame, supported the load, in this case the entire antenna structure. In operation, high-pressure oil pumped through vertical holes in the pad caused the pads to lift slightly off the runner as the oil exited between the very flat lower surface of the pad and the equally flat upper surface of the runner. Although the "liftoff" distance was very small, only a few thousandths of an inch, the thin film of oil was of sufficient strength to allow the three pads carrying the enormous load of the antenna to rotate freely in the azimuth direction around the circular track of the runner. One could imagine the antenna being supported by an infinite number of molecule-sized ball bearings rolling between the pads and the runner.

As mentioned earlier, the three pads were tied together by a triangular metal framework. This formed the base of the alidade structure and was called, obviously, the base triangle. At the center of the base triangle, several sets of heavy roller bearings called trucks were forced against a steel band called the radial bearing that was attached to the center collar of the pedestal. As the base triangle with its load rotated around the hydrostatic bearing, the three trucks, pressed against the radial bearing, constrained the pads to track precisely around the runner of the hydrostatic bearing. Therein lay the answer to an obvious question: What kept the pads from sliding off the runner?

The third key element of the azimuth rotation mechanism was the bull gear. The bull gear consisted of a massive, toothed metal ring attached to the pedestal just above the runner, as noted in the antenna structure diagram, figure 6.2. The teeth on the bull ring engaged with toothed pinions—small gearwheels—on each of the four azimuth drive assemblies attached to the base triangle. When the azimuth drive motors were activated, the pinions dragged the base triangle and its load around the bull gear to impart the desired azimuth motion to the alidade.

Horace Phillips watched with particular interest as the heavy pieces were lifted into position. Yet another brilliant young engineer pirated by Merrick from the industrial sector, Phillips had joined the Hard Core Team at JPL early in 1962 to oversee the hydrostatic bearing design then being integrated into the Blaw-Knox one-year study. For more than two years he had been the team member most concerned with the installation and setting of the hydrostatic thrust bearing, and Merrick held him responsible for its design and successful implementation. Now Phillips was seeing the physical realization of his design concept being set gently in place on the pedestal.

After Bob Hall secured the prime construction contract for Rohr, he had subcontracted the design and construction of the hydrostatic bearing to the Rucker Company of Oakland, California. Rucker in turn subcontracted the basic design work to the Friction and Lubrication Laboratory of the Franklin Institute in Philadelphia. There the complex variables of fluid pressures and flows, oil film height, and deflection of pad and runners and concrete under load were analyzed using advanced computer programs and scale models. In the final design, Franklin developed the bearing configuration, including the dimensions of the pads and runners, and established a stiffness requirement for the supporting concrete that would match the deflection of the runner as each loaded pad moved around it. Rucker translated these numbers into detailed mechanical designs and drawings for the elements of the bearing and its associated hydraulics. The individual pieces were then fabricated locally by specialty manufacturing companies. It had taken many layers of highly specialized engineering expertise to turn Horace Phillips's original ideas into the hardware he watched being set in place that day in mid-1964.

The roar of big diesels drifting across the still evening air of the high desert was a familiar sound in those days—this was part of the Fort Irwin Military Training Center, and all manner of military vehicles, tanks, armored carriers, and big trucks frequently passed near the Goldstone work site on various training exercises. But this evening, although the sound was the same, its origin was different. Four heavy-duty Peterbilts were grinding their way up the long slope before making the short descent into the antenna construction area. From the storage area where Don Ramsey, Rohr's on-site manager, stood waiting, the flashing lights of the escort vehicles came into view as they crested the low range of hills that surrounded the antenna site. The headlights of the lead Peterbilt and its wide-body, low-load trailer appeared a short time later. Soon all four vehicles with their escorts were strung out across the desert in a brilliant parade of red, white, and yellow lights. The first of the big steel had arrived at Goldstone, and more would soon follow.

It had been a long trip from Paramount, near Los Angeles, where the massive steel beams that formed the alidade structure had been fabricated at Precision Fabricators Incorporated. Trucking the beams was not a simple task. The largest pieces were 40 feet long and weighed 50 tons. The vertical members were fabricated from four H-section lengths of special steel that were welded together with the flanges at right angles to form a square central box section of great

Figure 6.3. Alidade Beam Section. The composite of four H-sections is approximately 52 inches wide. The web in each H-section is about 2 inches thick and the flange 3 inches thick. All of the steel in the alidade structure was required to meet certain chemical and grain-size specifications to prevent catastrophic failure due to brittle fracture, a failure mode that can occur in common structural steels exposed to temperatures below freezing. NASA/JPL.

rigidity. They were rigid and they were heavy: every thirteen inches of 4–H beam weighed one ton. The photographs, figures 6.3 and 6.4, show the impressive size of the individual beams and the complexity of the assembled alidade structure.

When all the pieces arrived at Goldstone, the ironworkers began to assemble them on the pedestal. As Don McClure described the process, it all seemed quite straightforward. "Everything was bolted together. First, the beams that formed the base triangle were lifted up separately and bolted to the pad assemblies. Each pad assembly consisted of two parts: the rectangular pad with a spherical socket attached to the top of it, and a corner weldment piece with the matching ball for the socket attached to it. The corner weldment allowed you to connect the corners of the base triangle together as well as connect the vertical pieces that eventually carried the elevation bearings. So, with the three corner weldments in place, we lowered the beams between the corners and bolted them together to form the base triangle. Once the base triangle was developed, we put on the vertical members that went up to carry the elevation bearings" (McClure 1999a).

Figure 6.4. Side View of Alidade Trial Erection. Before the components were trucked to Goldstone, the entire alidade structure was erected at the fabrication plant, to verify that it could be assembled correctly on-site. The massive 4-H beams that form the structural members of the alidade are clearly visible. The flat surfaces at the top of each of the two vertical members are 57 feet above the base triangle and will eventually support the elevation bearings. One side of the base triangle can be seen at bottom left. NASA/JPL.

While ironworkers labored with cranes and big wrenches to reassemble the alidade structure, Rohr's other subcontractors pressed forward to complete the installation of power plant equipment, azimuth drive gear, servo systems, electrical supplies for motors, pumps, electronics, and utilities, buildings and control room services, and general facilities. Driven by Merrick's unforgiving master schedule, a tangible sense of urgency pervaded the construction site. JPL engineers from the Hard Core Team continuously monitored the progress of the work at Goldstone. In this environment the alidade structure was completed in late November 1964, and the stage was set for a most significant milestone in the progress of the project, the Alidade Trial Rotation. The completed alidade structure mounted on the pedestal is pictured in figure 6.5.

Alidade

"The day the alidade moved" was, up to that time, the major milestone date on Merrick's construction schedule, since it represented the first occasion on which two of the most critical parts of the overall antenna structure, the alidade and the azimuth bearing, would work together. There was considerable uncertainty about the outcome of the test. Understandably, the Rucker Company engineers were also apprehensive, since Rucker was the subcontractor responsible for the hydrostatic bearing, and it stood to lose heavily in money and reputation if the test failed. The design value for the film height, or clearance between the bottom of the pads and the top of the runner, was five thousandths of an inch, the thickness of a few sheets of computer paper. Any significant irregularities in the runner surface could cause the pads to "ground out" as they passed over them. The resultant damage to the finely polished surfaces would have devastating consequences. This was the main concern lurking in everyone's mind that day. They could not have known just how real that concern would become in the months and years ahead.

For the trial rotation, small electric motors were adequate to turn the alidade without the full load of the antenna upon it. They would be replaced with powerful hydraulic motors later. Special instrumentation had been attached to the pads to measure film height and rotation angle. The lead Rucker technician activated the hydraulic pumps and slowly raised the operating pressure to 2,500 pounds per square inch. An excited announcement of "liftoff," meaning all three pads had been elevated clear of the runner, was greeted with a hearty cheer from all hands. Then the Rucker technician applied power to the azimuth motors and, almost imperceptibly, the huge structure began to rotate about its azimuth axis. Everyone practically held their breath while it continued to move slowly, at one-tenth of a degree per second, for a full 30 degrees. Five long minutes

Figure 6.5. Advanced Antenna Construction at Goldstone, November 1964. Supported by the three hydrostatic bearings at the corners of the base triangle, the completed alidade structure sits atop the pedestal. The elevation bearings will eventually be placed on the flat bases at the top of the two vertical members of the alidade. NASA/JPL.

dragged by with no alarm indications, and the movement halted. The test was repeated in the opposite direction with similar results. A few days later, the film height was checked over the full range of azimuth rotation with equally satisfactory results. There were variations, of course, but the film height remained within the specified limits of five to eight thousandths of an inch. To say Merrick was pleased would be a gross understatement. Merrick always wore a big grin when he felt good, and that day the grin stretched literally from ear to ear. Back at Rohr, Bob Hall received news of the test with equal pleasure, and felt that his confidence in his choice of Rucker to build the hydrostatic bearing had been vindicated.

In his regular monthly report for December 1964, Merrick recorded the event rather impassively: "A major milestone was passed on December 4, when a successful hydrostatic 'liftoff' and azimuth rotation was accomplished. All three pads of the hydrostatic bearing, with the weight of the large alidade structure upon them, were lifted and floated. Rotation through 30 degrees was accomplished in a successful trial of the component. Further rotation occurred on December 5 and 6 when detailed measurements of the film thickness was verified during a 340 degree rotation" (Merrick 1965).

With the master schedule indicating "Azimuth Trial Rotation complete," attention turned to the other axis of rotation, the elevation or tipping axis. Rotation in elevation lay a long way ahead, and the first step toward that milestone involved two solid-steel monoliths called elevation bearings.

As the first year of construction drew to a close with the project on schedule and on budget, optimism ran high in Pasadena, San Diego, and Washington, D.C. All indications were that the AAS project would be brought to a successful conclusion just over a year later, as planned.

Although it started badly, it seemed 1964 had been a good year for JPL. *Ranger 6*, the fourth attempt to take close-up pictures of the lunar surface, had been dogged by a failure in the TV camera circuitry, although the spacecraft impacted the surface right on target. Learning quickly from that unfortunate experience, JPL engineers corrected the circuit problem and on the last day of July were rewarded with outstanding success. In the last fifteen minutes before impact, *Ranger 7* transmitted 4,316 close-up images of the lunar surface to the Goldstone 85-foot antenna. Scientists were more than satisfied with the results. Most of the images were as much as a thousand times better than Earth-based photos. It was concluded from the pictures that the mare regions of the lunar surface would be smooth enough for a later manned landing.

But there was more. *Mariner 3,* the first of JPL's missions to Mars, had experienced two severe failures shortly after launch that led to loss of contact with the ground network. Out of control, the spacecraft entered a useless heliocentric orbit. However, the launch of the second Mars mission, *Mariner 4,* in late November had been completely successful, and indications were that *Mariner 4* might well succeed in reaching Mars. While its encounter date in mid-1965 would be too early for the new antenna to take part, a success at Mars would be but a preamble to what would follow, and for that the Big Dish would be ready.

The Soviet program had, however, experienced a long and discouraging string of failures. Repeated attempts to reach the Moon had been unsuccessful, or only partially successful, as had several Venus missions and one mission to Mars (Siddiqi 2002).

For both programs, the travails of venturing into deep space were proving to be intimidating, to say the least, but overall NASA was slowly pulling ahead.

By 1965 Fred McLaughlin's association with the elevation bearings went back a few years, almost to the beginning of the project. It would extend past the end of the project for more years than he could possibly have imagined.

Here is McLaughlin (1999) describing his first acquaintance with Bill Merrick:

> I first became closely involved with Bill Merrick when I was at Goldstone as the Blaw-Knox engineer for the construction of the second 85-foot Az-El antenna late in 1959. Bill Merrick, understanding that there were larger things ahead on antennas, suggested that I might consider coming to work for him at JPL. When we completed that antenna in February 1960, I returned to Pittsburgh maintaining that I would never go to work for Bill Merrick. A few months later, in the summer of 1960, when I returned to Goldstone to deal with a repair problem on the 85-foot Az-El antenna, Bill Merrick started to pressure me because he had plans to go on to build a really giant antenna, and I began to think seriously about it. Finally I agreed, and in August 1960 I reported to JPL as a senior research engineer with Bill Merrick in his Antenna Structures and Optics Group. Work on the original specification, EPD 5, was just under way when I arrived.

Merrick had successfully pirated industry, in this case Blaw-Knox, for yet another addition to his Hard Core Team.

After the pressure-cooker atmosphere of industry, McLaughlin quickly adapted to the low-key, campuslike environment that prevailed at JPL. He recalled the team spirit of the time: "The team spirit was very real and the group referred to itself as the Hard Core Team from early on. Merrick really believed in getting people co-located—engineers, procurement, and administrators alike. It fostered the team spirit and made for better informal communications and faster response times" (McLaughlin 1999).

Was Merrick perhaps too highly focused on the project to care much about people? "I think people either connected with Bill or they didn't," McLaughlin observed. "If you connected, you were willing to work extremely hard for him and overlook his shortcomings. If you didn't connect, all you could see were his shortcomings. There were times of personal difficulties when Bill claimed to be understanding and would offer any help that was needed, but then he seemed to disconnect from the relationship and his only acceptable focus would be on the work" (McLaughlin 1999).

Was there ever any question about his technical ability, judgment or leadership? "No, never. However, Bill was not the kind of person you would recognize as a leader of men. He led by example and by being the most important member of a dedicated group of followers. He herded us rather than led us, he was a herder of men rather than a leader." Later McLaughlin added, "But in defense of his leadership qualities, there was no need for him to give the appearance of a leader since he had the quiet competence of Bob Stevens and the outstanding leadership of Eb Rechtin at his disposal" (McLaughlin 1999).

McLaughlin became Merrick's cognizant engineer for the elevation bearings and the gears for the AAS project. He was an excellent draughtsman, an attribute much admired by Merrick, and he had acquired considerable experience in his areas of responsibility through his work on the Blaw-Knox antennas.

The first of the two giant elevation-bearing assemblies arrived at Goldstone on November 2. The special tractor and loader then returned to San Diego and returned a week later with the second one. The two assemblies were fabricated by the National Steel and Shipbuilding Company under subcontract to Rohr. The Swedish company SKF supplied the spherical roller bearings, and Bethlehem Steel Company in Pennsylvania made the huge castings that held the whole assembly together. Figure 6.6 shows one of the elevation-bearing assemblies being delivered to the work site.

At first sight, the elevation bearings just looked big, but they were in fact much more complex than they looked. Within each of the two bearing assemblies, two spherical roller bearings manufactured by SKF supported the 25-inch-diameter shaft that carried the full weight of the antenna reflector and its associated tipping parts, 3 million pounds in all. SKF was intimately involved in the design and construction of the entire assembly. This decision, too, would prove to be of immeasurable importance in dealing with the unforeseen events that took place in the years ahead.

McLaughlin also worried about the complex hydraulic-mechanical devices called drives that imparted azimuth and elevation motion to the antenna. The azimuth and elevation drives were similar in design, although they used different structural mounting arrangements. There were four drives for each axis. On

Figure 6.6. Delivering Elevation Bearing Assembly to Goldstone. The elevation bearing was trucked to the work site from San Diego on a specially strengthened loader. Approximately 14 × 11 × 10 feet and weighing 65 tons, it was the heaviest single piece to be shipped to the site during construction. The two elevation assemblies were mounted on the top of the alidade and were tied together by a horizontal tie-truss that attached to the large flat mounting plate seen at the top of the upper bearing casting. Later the reflector structure would be attached to the tie-truss. NASA/JPL.

each drive, a hydraulic motor drove the input shaft of a reducer (gearbox) that had a 20-inch pinion (gearwheel) attached to its output shaft. The pinion meshed with a circular segment of toothed track called a bull gear through which the drives imparted motion to the elevation or azimuth sections of the overall antenna structure.

Great care had been taken to ensure that the reducers and drive motors would meet the specifications for stiffness and backlash. Both of these parameters would contribute in a critical way to the overall pointing capability of the completed antenna. To verify that they did indeed comply, an extensive set of stiffness and backlash tests was performed in Pennsylvania before the units were shipped to Goldstone.

When the elevation drives and bull gear arrived on-site in February 1965, the two elevation bearings were already in place atop the alidade, and Rohr was preparing to hoist the tie-truss into position to tie them together. The two elevation bearings are clearly visible in figure 6.7.

Just as the alidade and azimuth bearing carried everything that rotated in the azimuth direction, the tie-truss and elevation bearings carried everything that rotated in the elevation direction. These were known collectively as the "tipping parts," as distinct from the "rotating parts" that were carried by the azimuth bearing. The tie-truss had to perform two important functions. First, it had to support the full weight of the tipping structure, including its own weight, without bending or flexing excessively. Second, it had to prevent the two elevation bearings and the supporting alidade structure from being spread or forced apart by the weight of the tipping parts and additional load due to the force of the wind. The tie-truss was the single structural member that transferred all these forces to the azimuth bearing and thence to the pedestal and its foundation.

But there was another important function performed by the tie-truss. It carried the two elevation wheels. Having two elevation wheels allowed the centrally located instrument tower with its top-mounted master equatorial instrument (ME) to pass between the wheels as the antenna tipped on its elevation axis. The instrument tower and ME were thus physically isolated from the movement of the antenna structure. The elevation wheels served primarily to impart the tipping motion to the antenna structure by means of a bull gear bolted to the rim of each wheel and one of Fred McLaughlin's elevation reducers mounted firmly on the alidade. They also carried the heavy lead counterweights that balanced the reflector and backup structure about the elevation axis.

To perform these functions properly, the tie-truss had to be strong and stiff. It was both, and that meant big and heavy. Figure 6.8, a photograph taken during a test run at the Rohr assembly facility, gives an impression of the great size of the combined structure.

Figure 6.7. Placing Elevation Bearing on Alidade Structure, January 1965. The tower derrick is placing the second of two elevation bearings on the alidade structure. Later they were tied together by a tie-truss that was laid across them and permanently attached at each end. The vertical tubelike structure is the wind shield for the upper portion of the instrument tower. This would support a small cabin (astrodome) to accommodate the master equatorial instrument, a key element of the precision optical pointing system. NASA/JPL.

Figure 6.8. Tie-Truss and Elevation Wheel in Test Assembly Fixture. The tie-truss with elevation wheels attached is mounted on a test fixture that emulates the actual elevation bearings. The tie-truss was 57 feet long, 12 feet across the base, and 6 feet in height. NASA/JPL.

Moving such a large structure presented its own special problems, which involved numerous road closures to allow the very wide load to pass. It took a week for two low-loader trailers, secured side by side, to move it to Goldstone.

After the elevation bearings had been carefully aligned and bolted down, the tie-truss was lifted in one piece and set end-to-end across them, as figure 6.9 shows. Don McClure made some precision measurements to verify that the sag at the center of the 57-foot (17.4-meter) span between the bearings satisfied Merrick's specifications. Then, confident that the tie-truss would perform correctly with the full load of the tipping parts upon it, Rohr began to assemble the

Figure 6.9. Tie-Truss Being Placed on Elevation Bearings, May 1965. After the end weldments of the tie-truss were bolted to the elevation bearings, the two elevation wheels, one on each side of the instrument tower, were added. The upper part of the instrument tower is clearly visible between the vertical members of the alidade structure. Most of the lower portion of the alidade had now been enclosed to accommodate hydraulic, electrical, and mechanical equipment associated with the servo drive system. NASA/JPL.

antenna elevation wheels and backup structure. The big-steel and heavy-engineering phases of the construction job were complete, and on schedule.

To this point in the project, the primary objective had been to create an extremely stable structure that could carry a load of 2.6 million pounds, could be moved with great precision in azimuth or elevation, and would maintain those capabilities in wind speeds up to 70 miles per hour. In a stowed position it would survive winds to 120 miles per hour. To that end, the five major elements of the nontipping part of the overall antenna—the pedestal, azimuth bearing, alidade, elevation bearings, and tie-truss—had been designed for stiffness, or rigidity, rather than strength. Referring to the alidade, the final report said: "The design problem was unusual in a large structure, in that deflections, rather than stresses, were the governing criteria. In general the member stresses in the alidade were 25 to 30 percent of those normally encountered in large structures" (JPL 1974).

Although it was not yet apparent, the overarching requirement to meet the stiffness criteria brought with it a surplus of weight-carrying capability that would prove to be of inestimable value twenty years later, as we shall see.

Merrick was pleased, perhaps elated, as the focus of the Hard Core Team effort moved from heavy engineering toward the reflector structure and into the field of microwave engineering, a technology in which JPL was much more experienced.

Big Dish

Even before he won the contract, Bob Hall had decided that responsibility for the reflector and for the critical backup structure, the framework that held it all together and provided the means by which it was attached to the tie-truss, would remain at Rohr. He carefully selected subcontractors to perform the work in all other areas, but he retained the reflector fabrication for Rohr. It was the kind of work that Rohr did best. How much of the antenna was actually built at the Rohr plant? Hall's reply (2000) was very specific:

> We built the total stretch-formed reflector [panels] and the accurate backup structure. That's what I went there for. We used the Slab[1] to fabricate the accurate parts, like the backup structure and fitting the tie-truss to the elevation wheels. In other words, the more accurate we built the parts, the smaller the tolerances we would have to allow for in our [on-site] adjustments. How accurately you control the front chords, and how accurate you make all those truss members that connect to them, determines that the point where you are going to connect a panel is in the right place.

> In a 210-foot structure, there are so many tolerances that add up that you must have accurate fabrication of the many parts. You don't care quite so much about the alidade as long as it is strong and stiff and it has the right attachment points. But if you get the backup structure up there [on the tie-truss] and you are a couple of inches off, what are you going to do? In normal structural practice that would be just great. But not for us. We needed to be within half an inch, so that it would be within range of our adjustment. So we did the precision fabrication. That is what Rohr did best.

The tipping part of the antenna was made up of six major components: the reflector backup structure, the parabolic primary reflector surface, the hyperbolic secondary reflector, the intermediate reference structure, the Cassegrain feed cone, and a quadripod support structure. With the exception of the secondary reflector, which was supported by the quadripod, everything was directly supported by the backup structure, as figure 6.10 shows.

The reflector surface was supported by a parabolic space frame 210 feet (64 meters) in diameter that was constructed from a network of 48 rib trusses radiating outward from a central hub. The rib trusses were interconnected by a rectangular girder, several circular hoops, and various other structural members. The rectangular girder provided the framework by which the reflector was attached to the tie-truss.

The surface of the primary reflector comprised 552 individual panels contoured from aluminum sheet and riveted to the precisely formed space frame. The individual panels were attached to the reflector backup structure with adjusting screws that were accessible from the exposed front surface of the reflector. All of the panels over the outer half-radius were perforated to reduce the effective weight of the huge reflector.

Because Merrick's team had elected to go with a Cassegrain type of antenna design,[2] they needed a hyperbolic-shaped secondary reflector. This subreflector was supported by a four-legged structure known as a quadripod. Each of its four legs was attached to the rectangular girder at one end, and all four came together to support the subreflector at the other. Rigidity was a key feature of the quadripod. It was essential that the subreflector maintain its position relative to the main reflecting surface while the antenna tipped from horizontal to vertical, regardless of the wind conditions. Any relative movement would "defocus" the radio beam, or skew it to one side, or both.

The subreflector was a hyperbolic-shaped precision radio-reflecting surface 20 feet (6.1 meters) in diameter, with a short skirt to improve its "appearance" to microwave radiation. Remotely controlled electric motors and screw jacks at-

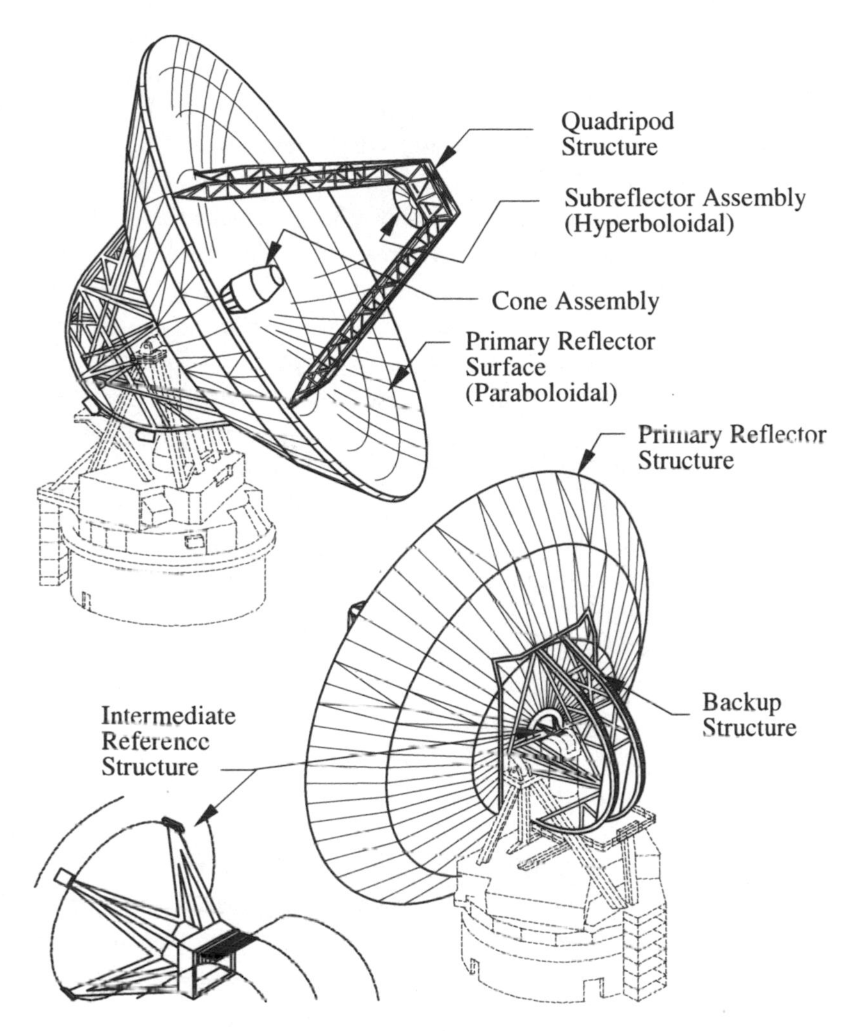

Figure 6.10. Antenna Tipping Parts. The diagram identifies the major components of the tipping parts of the antenna, all of which are supported by the tie-truss. Illustration by Fred D. McLaughlin, NASA/JPL.

tached to a framework behind the reflecting surface enabled some degree of adjustment for optimizing the geometry of the microwave feed system.

In the adaptation of the Cassegrain principle to the radio antenna for space communications, microwave energy collected by the primary reflector was first concentrated onto the surface of the subreflector. The subreflector redirected the energy to a convenient focus point near the center of the primary reflector. A conical microwave horn placed at that point fed the collected energy directly to an ultrasensitive radio receiver (Potter 1962). The microwave horn, ultrasensitive receiver,[3] and ancillary electronics were contained in a tall, conical housing that was mounted at the center of the main reflector and supported by the hub.

The so-called intermediate reference structure (see figure 6.15) defined the true direction of the radio frequency axis of the reflector surface—that is, the true direction of the radio beam. It was a rigid, weblike structure centered on the reflector boresight axis and attached to the reflector center hub. A small optical mirror carried at its apex formed the reference surface that, in conjunction with the master equatorial instrument, comprised the precision angle data assembly.

For the most part, Bob Hall depended on his site manager Don Ramsey to ensure that the work at Goldstone progressed smoothly and stayed on schedule. He had chosen his subcontractors very carefully and had high confidence in their ability to deliver their products as specified, on schedule and in budget. This permitted him to devote much of his attention to the fabrication, assembly, and testing of the high-precision tipping parts of the antenna described above.

He had persuaded Pappy Rohr that he needed a place to lay out and assemble all the pieces. He referred to it as the Slab and used it extensively for that purpose. There was room for accurately positioned welding jigs and for assembling, testing, and checking the larger components. With the exception of the reflector surface panels and the subreflector, all the components that Rohr supplied passed through the Slab. At any particular time, the state of the Slab reflected the progress of on-the-job construction in the distant Mojave Desert.

As the components were completed, they were trucked to Goldstone and assembled in accordance with Rohr's instructions. Suddenly the outward appearance of the antenna began to change rapidly. Whereas there had been little observable progress over the past year, now the antenna seemed to change daily as the big tipping parts were hoisted into position. By May of 1965 the rectangular girder and elevation wheels were in place, as shown in figure 6.11.

With the main girder and central hub in place, the Rohr workers began attaching the ribs that would eventually form the inverted umbrellalike structure to which the surface panels would be attached. Attaching the ribs was not a job for height-intimidated workers. That task required high-steel ironworkers, and Rohr brought the best that could be found to Goldstone for the job. Sections of the outer ribs were assembled on the ground before the 300-foot-high derrick crane lifted them into position for attachment to the intermediate ribs that supported the inner half of the reflector structure. They made rapid progress, and by mid-May the intermediate rib cage extended out to the 56-foot radius of the reflector surface.

Don McClure (1999a) described the way they went about attaching the outer ribs. "Once the inner ribs were in place, we started attaching the outer ribs. These were all preassembled and lifted up in groups of three or six. As each

Figure 6.11. Tie-Truss, Elevation Wheels, and Rectangular Girder, May 1965. The rectangular girder and central hub upon which the reflector will be built have been completed. The two semicircular elevation wheels hang below the tie-truss in this photograph. Large elevation "bull gears" would be attached to the periphery of the wheels later. NASA/JPL.

group of ribs was attached, the antenna was rotated [in azimuth] to keep the antenna balanced. When we attached one group of three, then we would rotate the antenna and put a group of three on the other side. We kept rotating it until all the ribs were installed." But there was more to it than merely bolting on the ribs. Each one had to be meticulously aligned before it was permanently fixed in place, so that the completed structure would form a near-perfect parabola to which the surface panels could be attached later. McClure continued, "We had a theodolite in the center of the dish, and there were targets on the ribs so that we could set them to the right height and the right location. We set the theodolite to a predetermined angle and adjusted each rib until the target sat on the theodolite crosshair. Then we tightened up all the bolts. There were no problems with any of that."

As the workers became accustomed to the process, the work accelerated. By the beginning of summer 1965, the backup structure for the reflector surface was nearly complete, as the aerial photograph shows in figure 6.12.

The Rohr workers began to install the reflector surface panels as soon as the backup structure was complete. The reflector surface was composed of 552 aluminum panels that extended from a 20.5-foot-diameter center hub to the outer edge. They were arranged in eight circular rows, the panels of the outer four being perforated to reduce weight and wind-loading effects, while the panels of the inner four rows were solid surface. The number of panels in each row varied, the outer rows having many more panels than the inner rows. Each row of panels had its own particular shape, but the panels within each row were the same shape. Rohr fabricated the panels from an aluminum skin that was stretch-formed to its unique shape and riveted to an aluminum frame formed from zee members that were also stretch-formed to an identical shape.[4] Each panel was carefully checked for surface accuracy to determine its deviation from a parabolic shape before being treated with specially formulated white paint to minimize solar heat absorption and shipped to Goldstone. Individual panels varied in length (about 10 to 14 feet) and width (3 to 6 feet), tapering slightly from top to bottom. They weighed 75 to 150 pounds each and were fastened to the backup structure with an adjustable screw arrangement that enabled each panel to be individually set to the correct position on the parabolic surface. Again Don McClure explained how those very critical adjustments were made. "To adjust the panels we had a theodolite in the center of the antenna and a small target that snapped into a little hole right next to the adjustment keyway. Each panel was just screwed up or down to get the right theodolite angle setting. Most of the adjustments were made at night to avoid thermal stress and to keep all the panels at the same temperature. And then we adjusted the other three corners around it to feather into it. Then we went to the next corner and set it and

Figure 6.12. Aerial View of Reflector Backup Structure Nearing Completion, July 1965. Most of the outer, cantilever ribs have been erected in this view of the antenna. The quadripod legs have been attached to the corners of the rectangular girder that is clearly visible through the maze of the inner rib cage. The central hub can also be seen inside the rectangular girder. On the ground, the last several preassembled outer ribs wait to be hoisted into place to complete the circular backup structure. NASA/JPL.

feathered those three. Then the screws were locked down" (McClure 1999a). Figure 6.13 shows the panels being installed on the antenna surface.

As figure 6.12 shows, the surface panels were mounted on the supporting backup structure in daylight with the antenna in its zenith, or vertical pointing, position. This was simply for safety reasons and for ease of working conditions. The actual adjustments for surface alignment were done with the antenna set at an elevation angle of 45 degrees, the "rigging angle" at which the distortions owing to gravity were optimal for the overall structure. To avoid unbalanced thermal effects, the delicate surveying adjustments were carried out at night with little or no wind. The surveyors used a special "all attitude" theodolite that

Figure 6.13. Installation of Reflector Surface Panels. The photograph shows the four inner rows of solid panels and two of the four outer rows, which are 50 percent porous. Two of the quadripod legs can be seen to the top right, and the hole for the Cassegrain feed cone support structure is in the center. A striped umbrella shades the theodolite instrument at the center of the picture. Note: Most of the critical surface observations were taken at night to minimize disruptive thermal effects. NASA/JPL.

was mounted at the vertex of the paraboloid to sight on small targets attached to each panel, as McClure explained. When Rohr's alignment work was completed, the operational surface accuracy of the reflector surface was found to satisfy the specified requirements in all respects and to exceed them in most cases.

The actual numbers were meaningful only in the context of the full set of complex conditions that the antenna was called upon to satisfy. However, an impression of the excellence of the design and its implementation was conveyed by the values for the operational reflector surface accuracy: specified value 0.250 inches (6.350 millimeters), measured value 0.170 inches (4.318 millimeters). Although he had not quite reached the one-eighth-inch mark, Bob Hall had exceeded the original design goal by a handsome margin. He was well satisfied, and so was Merrick.

The erection of the backup structure and surface panels progressed smoothly until August 1965, when a serious incident occurred while the antenna was being rotated in azimuth to place the last of the ribs in position. Don McClure was there, and remembered it well. "One day around this time when we were rotating the antenna to place some ribs, there was a horrible screech and everything ground to a halt. It was obvious that something serious had happened. One of the pads had grounded. Everybody came running out to see what had happened. So we drained all of the oil out of the reservoir surrounding the runner—that's a lot of oil—and, lo and behold, the top of the runners had these big scratches and gouges running most of the way round" (McClure 1999a).

Close examination revealed that the pad had been gouging the runner for some time, gradually worsening until the pad finally grounded out on the runner surface. Using time-sequenced photos of the job site to match the antenna motion with the work activity log, McClure and Phillips were able to determine the date on which the trouble first occurred. It corresponded with the date that a technician had reset the oil film height on the hydrostatic bearing hydraulic system. Obviously the film height had been set too low for the additional load on the alidade created by the components that had been added since the initial settings.

Urgent action was required. Using strong lift-points that had been designed into the alidade for just that purpose, the complete alidade and reflector were jacked up—a load that by this time was about 5 million pounds—and the pads were removed for resurfacing. While that work was being carried out in a machine shop, the gouges in the runner were repaired *in situ*, and the surface was reground to the requisite flatness. Eventually the pads were replaced and the system restored to service. The whole process took about one month. Referring to this episode, the final report said: "It was a grim reminder of the importance of incorporating methods for repair of major mechanical components into the design of the facility" (McClure 1999a).

By August 1965, with all of its prime reflector panels in place, the antenna had taken on its ultimate form and substance and, to the casual observer, appeared to be complete. Despite appearances, a lot more remained to be done. Externally, the master equatorial instrument had yet to be mounted atop the instrument tower and the subreflector and feed support cone placed in position above the prime reflector surface. Internally, technicians had not yet finished the installation of servos, computers, recorders, receivers, and ancillary electronics and services.

In addition to providing very stable support for the subreflector, the quadripod served as an ingenious lifting device for placing the Cassegrain feed cone structure in the center of the dish. When the antenna was tipped to point horizontally, the quadripod apex acted like the horizontal boom of a crane, to which the load was attached by the hoisting gear and a suitable bridle. Once the load was clear of the ground, a gentle elevation movement of the antenna to the vertical position enabled the load to be placed in the desired position at the center of the dish. The subreflector, feed cone, and microwave equipment were hoisted into position in this manner.

Master Equatorial

It was one thing to point an antenna in a given direction and quite another to know precisely where the actual radio beam formed by the antenna was pointing. Merrick understood that fact very well. So did all the members of his Hard Core Team, though none better appreciated the difficulty of solving that problem than did Houston D. McGinness. Merrick had assigned to him responsibility for the analysis, design, and development of the overall servo system and master equatorial instrument. McGinness joined Merrick's team in 1963 just as the design work for the AAS was getting under way. He brought with him a wealth of experience in precision instrumentation systems from his work in the Southern California Wind Tunnel facility that had been part of JPL for several years. He had also been involved in developing precision bearing devices for military applications in Bob Stevens's guidance and control section before JPL became a NASA center (McGinness 1992).

The key to the precision pointing and angle determination capability of the AAS lay in a complex electro-mechanical-optical system that involved three major elements: first, the stable and independently founded master equatorial instrument (ME) that was mounted at the intersection of the azimuth and elevation axes of the antenna; second, the intermediate reference structure, which was designed to accurately represent the orientation of the radio beam formed by the reflector surface; and third, the precision servo-optical link that related the angular orientation of the two elements with each other.[5]

The ME responded to pointing instructions by reorienting its mirror to the desired coordinates. Sensing the angular misalignment between the new ME position and the original antenna pointing position, the optical autocollimating system sent an error-correcting signal to the azimuth and elevation servo systems. The servos responded by activating the azimuth and elevation (rotate and tip) drive motors and reducers to drive the antenna reflector in the appropriate direction to zero out the error. When the two mirrors were again aligned, the antenna stopped. By programming the ME with information describing the predicted path of an interplanetary spacecraft, the radio beam of the antenna could be made to track the spacecraft continuously to maintain a two-way radio link for navigation, telemetry, and command purposes, as desired.

This idea had been developed some years earlier in Australia for use with the 210-foot radio-astronomy antenna at Parkes. The Australian engineers had generously made all their design data available to JPL, and the challenge for McGinness was to build upon that design to create an even better system for the Goldstone 210-foot antenna.

McGinness developed some tight specifications for the ME, and Rohr in turn passed them on to a small optical instrument company in Pasadena called Boller and Chivens. They may have been small, but they did good work. When the accuracy of the completed instrument was measured, it came in at 2 to 3 arcseconds,[6] less than half the error that McGinness's specifications allowed. The master equatorial instrument undergoing laboratory testing is pictured in figure 6.14.

When the testing at Boller and Chivens was complete and McGinness approved the test results, the instrument was mounted in a large crate, made shockproof to eliminate any vibration that might damage its very sensitive bearings, and carefully moved to Goldstone. To minimize the handling, a crane was standing by to lift it into position on the instrument tower as soon as it arrived on-site. The lift and placement were carried out with extreme care, and not without some anxiety on the contractor's part, since it was still their responsibility. Bob Hall (2000) recalled the occasion. "The Boller and Chivens ME was a very delicate instrument, and irreplaceable at the time. The day we put it up [on the instrument tower] I took out a one-million-dollar, one-day insurance policy to protect the ME while we erected it with the crane."

Finally the ME was integrated with the other components of the precision angle data system and, with the servo drive system in operation, the antenna could at last be steered in elevation and azimuth from the servo console in the main control room.

In the diagram, figure 6.15, the ME is shown in position on the instrument tower, with the intermediate reference structure and mirrors of the optical angle data system attached to the central hub of the reflector surface. The ME was positioned so that the intersection of its axes coincided precisely with the intersection of the azimuth and elevation axes of the antenna.

By the beginning of 1966, the antenna was a 210-foot-diameter mechanical masterpiece. The antenna surface was as perfect as the surveying instruments could determine, and it could be steered manually or by computer from the central control room. The precision angle data system was performing well, and the azimuth hydrostatic bearing, alidade, and pedestal appeared to be stable and

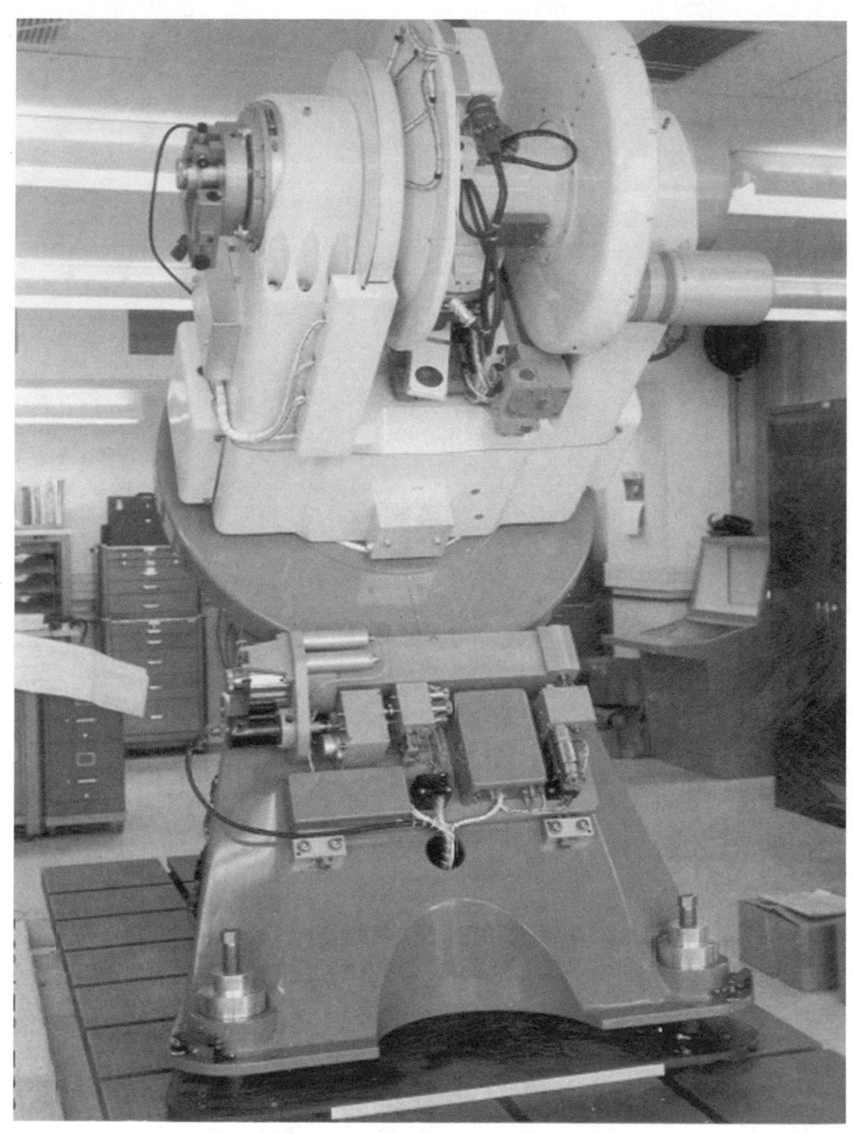

Figure 6.14. Master Equatorial Instrument. The master equatorial instrument was mounted in a dome-shaped enclosure (astrodome) on top of the instrument tower. The ME stood about 6 feet tall and 3 feet wide and weighed 4,000 pounds. NASA/JPL.

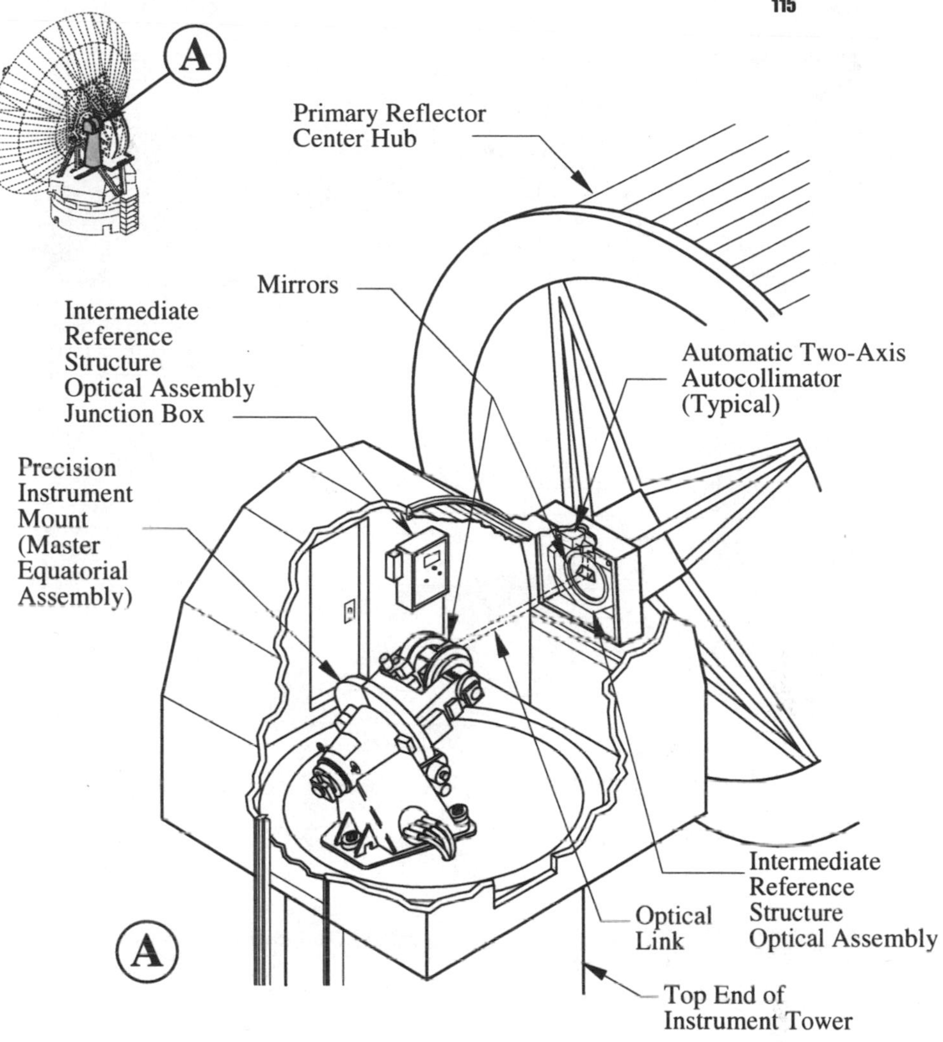

Figure 6.15. Master Equatorial Instrument and Components of Angle Data System. The ME used a polar type of mount similar to that used by astronomical telescopes, so it generated its angular coordinates in units of hour angle and declination. The optical autocollimation arrangement allowed the angle data assembly to make the coordinate transformation required to generate drive signals for the azimuth and elevation servos of the antenna. Illustration by Fred D. McLaughlin, NASA/JPL.

to possess the requisite "stiffness" to satisfy Merrick's specifications. To cap it all, the project was on schedule and on budget. Insofar as construction of the antenna was concerned, the responsibilities of the Hard Core Team were essentially finished. The contractor still had a few facilities to complete and some cleanup work to do, but the antenna per se was complete. Figure 6.16 shows the antenna in "as built" configuration at the end of January 1966.

Figure 6.16. As-Built Configuration of Goldstone 210-Foot Antenna, January 1966. The antenna structure was complete as far as the Rohr contract was concerned. Later JPL added a microwave feed cone to the top of the feed support structure and completed the internal microwave installations. At that point the antenna became a "microwave instrument" capable of receiving radio signals from distant spacecraft. NASA/JPL.

At the end, it all really came down to this: Given that Don McClure's concrete was hard enough and Ron Casperson's steel was strong enough, and assuming that the oil film in Horace Phillips's hydrostatic bearing was thick enough and the backlash in Fred McLaughlin's reducers was small enough, then, if the precision of Houston McGinness's master equatorial was fine enough and the parabolic surface of Bob Hall's reflector was true enough and if Bob Wallace's servos pointed well enough—then, and only then, would Bill Merrick's antenna perform as a complete system to the level required by its ambitious specifications.

Were they? The answer would not become clear until the antenna could begin to perform the precise radio frequency functions for which it had been designed and constructed. For that purpose it required the addition of a microwave feed, a supersensitive microwave receiver, and ancillary microwave components.

Microwave

Barely noticeable from a distance, and for the most part hidden within the feedcone support structure, the feed and its microwave components were a superb example of microwave technology and design. They were not, however, part of the Rohr contract. They were to be supplied by JPL, and JPL was well qualified to do that work. JPL, or more specifically Phil Potter's Antenna Microwave Group at JPL, had developed the Cassegrainian feed systems and microwave components for the 85-foot antennas then in operation around the network (Potter 1962).

Early in 1965 it was decided to take advantage of the proven capability of the existing family of designs and, at least initially, adapt one of the S-band cones originally designed for the 85-foot antennas for use on the 210-foot antenna.[7] The special feed cone that was to be used for testing the AAS was ready by the end of the year and was hoisted into position on the adapter assembly. The final act in the commissioning of the Big Dish, the actual receiving of radio signals from a distant spacecraft, was about to begin.

Potter had left JPL by then, and Gerry Levy, one of the original members of the Antenna Microwave Group, led the final act. While Levy was an excellent microwave engineer, he was at heart a radio scientist. It was the application of microwave engineering to the pursuit of scientific investigation that really appealed to him. Gerry Levy's calm, laconic temperament was a perfect foil for Merrick's obvious impatience for a showy demonstration of the new antenna's capability. Levy would not commit to a public demonstration until he was absolutely sure that it would be successful, and for that he needed to thoroughly understand the radio performance of the new instrument. He and his experts from JPL knew exactly how to go about it, and they wasted no time in getting started. Since the permanent S-band receiving equipment was still being installed in the control room, they set up a temporary facility in the alidade room, using a special receiver normally used for research and development at their field test station at Goldstone.

After the waveguide connections were made and all the necessary antenna test equipment was connected, there followed a lengthy series of evaluation and acceptance tests, each building on the previous one to correct discrepancies, make refinements, add more complexity, and penetrate deeper into the intricacies of the antenna's operation. The sharpness and purity of the radio beam was determined by measuring the strength of the signal from radio stars as they passed across its field of view. The antenna's collecting capability or "gain" was determined by comparing it to the gain of a standard microwave horn receiving the same signal. The main reflector surface panels were reset with much greater accuracy, and the subreflector was focused for optimal illumination of the reflector surface. Sensitive mechanical instrumentation attached to all the structural and mechanical parts of the antenna detected the minutest strains or changes in load on the components as the antenna rotated in azimuth and tipped in elevation. Particular attention was focused on the oil film height of the hydrostatic bearing as the antenna was rotated through its full range in azimuth. Under Levy's direction, measured performance was compared with predicted performance to the maximum possible extent. Careful analyses determined satisfactory explanations for any discrepancies. The engineers worked to laboratory-type accuracy, measuring to the hundredth part of a decibel (dB) of performance gain. After all, NASA was looking to a substantial performance improvement over its existing 85-foot antennas to justify its investment in the Big Dish. Obviously, performance accountability was serious work, and it was so regarded by Levy and his microwave engineers.

Eventually the microwave engineers were satisfied that they had the right answer, and announced a value of 61 dB for the measured gain of the 210-foot antenna at S-band. The measurement uncertainty amounted to a few tenths of a dB. In round figures, the 210-foot antenna had been determined to have a performance improvement over the existing 85-foot antennas of 8 dB.[8] It was, in nontechnical words, about six and a half times more "powerful" than its smaller contemporaries in the network. NASA was happy and JPL's telecommunications engineers were delighted. You could do a lot with that extra 8 dB of uplink and downlink capability. You could, for instance, reach the outer planets.

By the beginning of March 1966, Levy and his team were ready. The timing could hardly have been more fortuitous. Two distant planetary spacecraft, perfect radio targets for a spectacular performance demonstration of the new antenna, were immediately available.

Pioneer 6 had been launched into a heliocentric orbit in December 1965, and by March 1966 was about 28 million miles from Earth. Its eternal journey around the Sun was just beginning.[9] A steady stream of telemetry data at about 1,000 bits per second radiated toward Earth from its tiny 8-watt transmitter. The other spacecraft, *Mariner 4,* had been launched in November 1964. Now it too was following a heliocentric orbit that had taken it outside Earth's orbit to a flyby encounter with Mars in July 1965. As it passed within 6,000 miles of the Martian surface, it had provided the first close-range images of Mars and confirmed the existence of surface craters and, incidentally, confirmed the nonexistence of Martian canals. Following the Mars encounter, the spacecraft began pursuing a solar orbit that had been slightly modified by the effect of Mars's gravity field. As *Mariner 4* moved along its new path further and further away from Earth, the strength of its radio signal slowly decreased. By September 1965 the receivers on the 85-foot antennas could no longer distinguish the incredibly weak spacecraft signal from the background radio noise generated by their own receivers. *Mariner 4* was still working well and transmitting its science data to Earth over 191 million miles of deep space, but for the 85-foot antennas the spacecraft was history.

By March 1966 its distance from Earth exceeded 250 million miles, and the signal from its minuscule 10-watt radio transmitter was very weak indeed. It was approaching a point in its orbit called solar occultation.[10] Radio scientists already knew that the basic characteristics of radio waves were significantly disturbed by passage through the Sun's atmosphere—the solar corona—and that a great deal could be learned about the solar atmosphere by studying these perturbations to radio waves. By recording the radio signal from a spacecraft as it slowly entered occultation and again as it exited, radio scientists would have an unprecedented source of scientific data for later analysis. However, an antenna with a very sharp beam would be required to discern the fine details in the data. The new 210-foot antenna was just such an instrument. For Gerry Levy it was too good an opportunity to miss. But he knew that, irrespective of what happened on Earth, whether the engineers and radio scientists were ready or not, the spacecraft now moved inexorably toward its appointment with the Sun on March 17. There was no time to lose. They would first try to acquire *Pioneer 6*, since its signal was the stronger of the two, and tune up their equipment, then point the antenna toward the predicted position of *Mariner 4*. If they were successful on the first try, they would have a few days to refine the pointing coordinates and make further adjustments before the actual occultation sequence began.

On March 16 the great new antenna stood poised on the threshold of fulfilling its intended purpose. It was a unique moment in the history of deep space exploration. Later there would be a formal ceremony to mark the occasion, though not the instant. For that day, the record (JPL 1974) simply says, "The first reception of signals by the 210-foot antenna was on March 16, 1966 (Milestone 22), about 3 months less than 3 years from the signing of the contract with Rohr Corporation and 3 months more than 6 years from the initial serious study of the communications and tracking needs which the antenna was designed to satisfy. Within the accuracy provided by the initial measurements, which were made using a radio star, the gain of the antenna was greater than the minimum performance specifications, and the noise temperature was less than that specified."

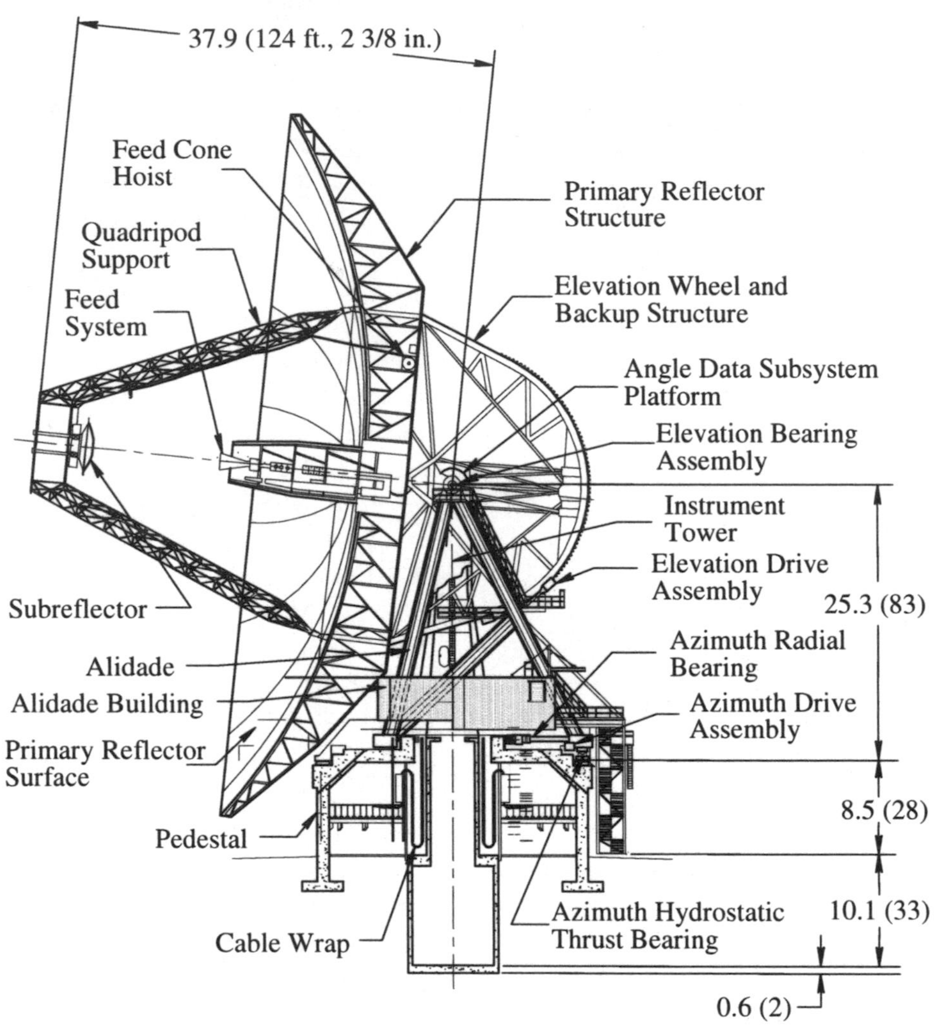

Figure 6.17. Major Components of Goldstone 210-Foot Antenna, March 1966. Illustration by Fred D. McLaughlin, NASA/JPL.

Regular daily tracking of *Mariner 4* began the following day. As the spacecraft neared solar occultation, it passed within 0.6° of the limb of the Sun, as viewed from Earth, and provided the first opportunity for scientists to observe the distortion of a radio signal passing through the solar atmosphere (Goldstein 1967, JPL 1974). Merrick wanted a spectacular public demonstration of what the new antenna could do, and now he had it.

Although these events marked the end of the construction project, they also marked the beginning of the operational life of the antenna. A great sense of euphoria prevailed at Goldstone, in Pasadena, in Washington, D.C., and of course at Rohr in San Diego. The great new antenna represented a remarkable cooperative effort between government, scientific institutions, and industry, and recognition was due to all who were involved in its success—the program administrators at NASA headquarters no less than the managers and designers at JPL or the construction engineers at Rohr and its subcontractors at Goldstone. It seemed appropriate to bring them all together to recognize their achievement.

And so a dedication ceremony and celebratory luncheon was planned for Goldstone on April 29, 1966. A press release was prepared, and the print news media were alerted. The news release presented the following interesting construction data (NASA 1966):

Diameter of "dish" reflector	210 feet
Height of antenna with dish pointed at horizon	234 feet
Circumference of dish	715 feet
Area of dish	0.85 acre
Height of pedestal	34 feet
Diameter of pedestal	83 feet
Depth of pedestal below ground	11 feet
Thickness of pedestal's reinforced concrete walls	42 inches
Central instrument tower, height above ground	73 feet
Central instrument tower, depth below ground	33 feet
Weight of entire structure	8,000 tons
Weight of reflector and alidade	2,500 tons
Weight of instrument tower	500 tons
Weight of pedestal	5,000 tons

> The alidade supports the reflector and its tipping mechanism. The alidade stands on three pads, each supporting 1,600,000 pounds, a total weight in the moving parts of nearly 5,000,000 pounds, approximately 2,500 tons. Each bearing pad floats on a film of oil (hydrostatic bearing) approximately 0.01 inch thick.

It was indeed a Big Dish. Its major components are illustrated in figure 6.17.

The Dedication

That spring morning in April 1966 dawned hard and clear over the Mojave. The desert floor, so barren and desiccated most of the year, was carpeted with a profusion of wildflowers, now refreshed by the cooling touch of the receding night. At Goldstone Dry Lake, utter silence pervaded the intimidating landscape in every direction.

By midmorning, several hundred guests and visitors had gathered in front of the horizon-pointed antenna at the Mars site. It was a splendid assemblage of prominent personalities from the world of science and technology. There were JPL director William Pickering and Caltech president Lee DuBridge; from NASA headquarters there were Robert C. Seamans Jr., deputy administrator, and Edmond C. Buckley, associate administrator for Tracking and Data Acquisition; and from Capitol Hill came Senator Clinton P. Anderson, chairman of the Senate Committee on Aeronautical and Space Sciences, and Congressman George P. Miller, chairman of the House Committee on Science and Astronautics. JPL managers and designers were well represented too. Among them were Rechtin, Stevens, and Merrick—and, of course, all of his Hard Core Team. Robert Hall and a senior executive represented the Rohr Corporation. Sadly, Pappy Rohr had died just a couple of months earlier. Most of the subcontractor firms were also represented.

The sun shone, the Fort Irwin Army Band played lively music, Dr. Pickering welcomed the distinguished guests and the officials from NASA headquarters, and the speeches began.

As the *Pasadena Star-News* reported (Swaim 1966),

> Edmond C. Buckley, NASA's associate administrator for tracking and data acquisition began by telling of "outstanding tests" already conducted by the giant antenna including the recording of a solar occultation experiment by the historic Mariner 4 spacecraft now more than 200 million miles away.
>
> Dr. Lee DuBridge, the president of Caltech, told why the facility was built at the Goldstone location. "In the old days of rocket research we moved the rocket plants out to the desert to take the noise away from the people. Now," he said, "we have put this antenna in the desert to keep the people noise away from it." DuBridge also told of a great new era of astronomy research, reaching farther than ever into the expanding universe with such marvelous mechanical and electronic achievements as the "210" which can study, without interference, the findings of flying observatories

> sent into space to do their work. Dr. Dubridge praised the Rohr Corporation of San Diego, the prime contractor on the project, and its thirty subcontractors. Among these was the William J. Moran Co. of Alhambra [California] which participated in the engineering phase of the project, particularly the problem-beset construction of the concrete base pedestal of the "210." The antenna, it was noted, had been so anchored as to withstand winds up to 150 miles per hour, and to operate during winds of 70 miles per hour.

At that point, as if to repudiate the assertions just made, a sudden gust of wind whipped the ceremonial red ribbon loose from the podium, sending it whirling into the air over the heads of the guests seated nearby. Embarrassed officials quickly secured the errant ribbon to a burst of applause, and the ceremony continued.

In his address, Robert C. Seamans Jr., said that without this great new antenna the signals from the *Pioneer 6* might soon have been lost. The signals had been getting weaker in their reception by the two existing 85-foot antennas at the Goldstone site. He also noted that the new facility would be ready to receive telemetry signals and television photos from JPL's Surveyor lunar soft landing missions, the first of which is scheduled for launch late in May. Seamans praised Dr. Pickering and the entire JPL staff for their great work. "The great event of this decade, if not this century, was the *Mariner 4* mission, which gave us not only new scientific knowledge, but close-up pictures of Mars. We are going so far out in space that spacecraft communications and tracking must be improved. What we have here is a quantum jump in that direction."

The *Star-News* concluded, "All speakers were in high praise of Dr. Eberhardt Rechtin, assistant JPL director for Tracking and Data Acquisition, under whose great leadership the entire Goldstone operation has progressed" (Swaim 1966).

After a brief address, Representative Miller cut the red ribbon and pressed a button on the podium to activate the giant new antenna.

Imperceptibly at first, and most improbably for such a huge structure, the Big Dish began to move. With infinite grace it tilted slowly upward until it pointed its invisible radio finger to the sky at about 60 degrees. Then it paused. For a moment the murmuring of the expectant crowd was suspended. A few flags snapped in the light breeze, a child worried, the tension grew. Suddenly the high-pitched, sharp-edged sound of a telemetry signal broke the silence. The Big Dish was receiving a radio signal from the *Pioneer 6* spacecraft, 28 million miles across the infinite space of the solar system and, like the Earth itself, traveling in an eternal orbit around the Sun. The crowd released its tension in a torrent of applause, cheers, and whistles, and the band struck up with appropriately heroic music. Gathered on the platform, the VIPs shook hands and congratulated each other.

As the *Star-News* put it, "The huge antenna dish tilted upward with surprising speed until it found its far-out target, the *Pioneer 6* spacecraft now orbiting the Sun, and picked up its signal loud and clear, from more than 28 million miles in space" (Swaim 1966).

For those familiar with the wonder of deep-space communications, it was a high-tech moment. For the others, it was a moment of high drama. For the engineers who had made it grow—from an idea in Merrick's mind, to a big hole in the ground and a small pile of wet concrete, to the magnificent structure that now stood before them—it was indeed a magic moment.

Ron Casperson was there, and he remembered (2000): "I was there in the audience with my wife. It was a beautiful warm day in the morning and I remember the wildflowers, they were eighteen inches tall, just rampant, the whole desert was just aglow, like they were celebrating for us. Then they announced that they going to try to acquire a signal from a spacecraft in orbit round the Sun. And then they rotated the antenna up to point to the sky. It moved up slowly just like a big beautiful flower. Suddenly comes this beeping over the loudspeakers. The whole audience just gasped. I had tears in my eyes. My wife reached over and gave me a great big hug. It was a magic moment for everybody."

Fred McLaughlin (1999) remembered it in a different way. He was aware of a potential problem in the elevation drive system that could bring the antenna to a grinding halt. The problem had not yet been corrected, and Fred desperately hoped it would not occur again on this most auspicious occasion. Just before the antenna started to move, one of his colleagues reminded him, "You know, Fred, by the end of the day you might be the most notable unemployed engineer in the country." He need not have worried. The problem did not recur, and a preventative fix was installed in due course.

The guests assembled for the dedication ceremony are shown in figure 6.18.

To the sound of triumphal music from the Fort Irwin Army Band, the ceremonies were drawn to a conclusion, and the antenna was opened for inspection while it continued to track *Pioneer 6*. NASA's newest facility, the Goldstone 210-foot-diameter antenna, the only one of its kind in the world, was officially in operation. By noon it was over, and the guests went their various ways. Some toured the antenna, some were taken to an official luncheon in the Goldstone cafeteria, some drove home to Barstow and Pasadena, and others simply returned to work at the site. Later the official party flew back to Pasadena, and the

Figure 6.18. Dedication Ceremony for Goldstone 210-Foot Antenna, April 29, 1966. The antenna was oriented to a zero elevation position to provide a backdrop against the desert landscape for the ceremonial proceedings. The lower edge of the Big Dish is visible across the top of the picture. After the ribbon cutting, the antenna moved upward to acquire and broadcast a signal from the *Pioneer 6* spacecraft, more than 28 million miles distant, orbiting the Sun. NASA/JPL.

Goldstone Dry Lake, its parched, cracked surface shimmering under the glare of the late afternoon sun, relapsed again into silence to await the coming night.

Largely unnoticed among the dispersing crowd that day was the loosely integrated group of about a dozen JPL engineers that had made it happen. All men in their late thirties and forties, all specialists in a wide variety of disciplines, they were Bill Merrick's Hard Core Team. They had the sole responsibility for bringing the 210-foot Goldstone antenna—or Advanced Antenna System, or AAS—into being. In this endeavor they were at various times, guided, directed, persuaded, cajoled, and inspired by their leader. In the end they were rewarded for their pains by seeing their exquisite engineering designs transformed into concrete and steel by an extraordinary engineer named Robert Hall, manager of the Antenna Division of the Rohr Corporation. The instrument of their creation now reached out beyond the limits of human perception to deep space where only spacecraft travel among the distant planets, and only astronomers heretofore had counted their movements. Their faith and confidence in their design and its transformation into reality, sustained over so many years, was truly vindicated that day, and each must have gone his way to celebrate according to his individual passion.

To Don McClure, Jet Propulsion Laboratory's "man in charge," this event carried an additional dimension of significance. For the past three years he had been responsible to Bill Merrick, the project manager, for overseeing all of the construction activity at Goldstone. The dedication ceremony marked the successful conclusion of that responsibility, and a turning point in McClure's career that, quite unexpectedly, would lead him back to Goldstone's Big Dish some twenty years later.

Within the next few weeks, Bill Merrick's engineers completed the contractual acceptance tests, and Bob Hall's people cleaned up the site and moved away. At the same time, JPL crews installed the electronic receiving and tracking equipment that enabled the station to begin spacecraft tracking on a regular basis. Finally, Merrick signed the documents that initiated the contractual transfer of the new antenna from the Rohr Corporation to NASA/JPL, and Merrick became the engineering custodian responsible for its well-being. Eventually he would transfer responsibility for operating and maintaining the antenna to the Network Operations organization at JPL. But for now it was under Merrick's control, and he exulted in that authority.

At the end of the contract, Rohr gave a big party in Pasadena for the subcontractors and JPL people that had been associated with the project. During the course of a reportedly boisterous evening, Bill Merrick was presented with a gold key to the control room of the new antenna, as a tongue-in-cheek symbol of a "gold-plated" contract successfully completed. Then Rohr quietly faded from the Goldstone/Pasadena scene. Don McClure returned to a staff position at JPL. NASA honored the Hard Core Team for its work on the AAS with the presentation of a Group Award at a special ceremony at NASA headquarters in Washington. Merrick was promoted to manage a newly formed Antenna Engineering Section at JPL and turned his attention to the construction of two similar 210-foot antennas, one in Australia, the other in Spain. Soon after, the team dissolved as a unique entity, although most of its members followed Merrick into his new section.

Figure 6.19. JPL Management with Model of the Big Dish, April 1966. Top-tier personnel at JPL admiring a table-top model of the 210-foot (64-meter) Advanced Antenna on its completion are, left to right: William D. Merrick, Robertson Stevens, William H. Bayley, Eberhardt Rechtin, William H. Pickering (director, JPL), Alvin R. Luedecke, Fred Felberg, and Walter K. Victor. NASA/JPL.

This change in responsibility marked a turning point in Bill Merrick's professional career at JPL. While the promotion to section manager brought with it commensurate rewards in salary and status, neither of these was of great significance to Merrick. He had found his greatest rewards in accepting challenges to his outstanding technical abilities, finding solutions to difficult problems and turning new ideas into real, working systems, as he had done with the AAS project. He could not find such a path in his new position. The new antennas in Spain and Australia, essentially replications of the Goldstone antenna, were to be implemented by others. His responsibilities now involved directing, managing, organizing, and interacting with highly qualified professional people at all levels to meet tight budgetary and schedule goals that were not under his control. Responsibilities such as these challanged his "people skills" but made little call upon his remarkable talents for creative engineering.

All of that, however, lay far in the future. It was now May 1966, Bill Merrick bore most of the credit and all of the engineering responsibility for the great new Goldstone antenna, and the Big Dish was about to become operational. Reporting the event a few days later, *Missiles and Rockets*, the premier aerospace journal of the day, said, "A 210-ft steerable antenna that will set the pattern for the United States Deep Space Network . . . over the next ten years, will become operational within the next two months" (Pay 1966).

The glory days for the new antenna were just ahead.

7 Soon There Were Three

Big Dish at Work

In addition to providing an exceedingly stable and rigid support for the moving parts of the antenna, the 42-inch-thick outer wall of the pedestal enclosed a large inner space that accommodated electrical and hydraulic machinery, workspace for the operating staff, some limited office space, and the main control room.

Tom Potter (no relation to Phil Potter) moved into the station manager's office on the lower floor in March of 1965 and began coordinating the installation of dozens of racks of electronic receiving, recording, data processing, and communications and control equipment that transformed the unimaginably small radio signals gathered by the antenna into intelligible, readable data for scientific and engineering interpretation and analysis. Potter had been at the Pioneer 85-foot antenna station at Goldstone for several years before he was assigned to the new 210-foot station, code-named Deep Space Station 14 (DSS-14). He was well experienced in deep space operations and possessed exactly the right personal disposition to deal with the myriad problems of managing the new station and the staff of technicians and operators to run it.

Although the S-band installation work was well advanced at the time of the dedication, it was not complete, and it was not until mid-1966 that Tom Potter was ready to begin limited support for in-flight missions from the new control room. He began with limited tracking support for NASA's lunar spacecraft—the Lunar Orbiters then orbiting the Moon, and the Surveyors then operating from the surface of the Moon. The two planetary missions, *Pioneer 6* and *Mariner 4,* also clamored for tracking time, since they were then well beyond the range of the 85-foot stations. To cope with the ever-increasing demand for tracking time on the Big Dish, Tom Potter was soon forced to increase his level of support. So began a trend toward oversubscription of antenna resources that would vex the network management for the next thirty years. It seemed that demand always expanded to consume whatever resources became available. Despite all the planning, in the real world there were simply never enough antennas.

Shimming Out of Trouble

Tom Potter pushed ahead with his station readiness, and by early 1967 he had DSS-14 running twenty-four hours a day, seven days a week. But then, at the worst possible time, he ran into serious trouble. The huge azimuth hydrostatic bearing, over which so much painstaking care had been lavished during construction, began to fail. At several points through its range of azimuth travel, the antenna triggered the film height alarm, indicating that the oil film between the pad and the runner had fallen below the safe minimum value and the pad was in danger of "grounding out" on the runner. The consequences would be drastic and could put the antenna out of action for a considerable time. In just a few months the *Mariner 5* spacecraft was due for its encounter with Venus, the first major event for which the new antenna would be the prime station. This situation, which would be the first of many, rapidly became critical.

Merrick was distraught. The antenna was still under his engineering cognizance, and he was responsible for any problems that developed until it was transferred to operational status. Ron Casperson (2000) recalled the desperate situation very clearly. "When the hydrostatic bearing failed for the first time, Bill Merrick took it very personally. He worked alone, virtually in hibernation, in a motel room in Barstow trying to figure out what to do about the bearing groundings."

It was not long, however, before Merrick and Horace Phillips developed a plan. They proposed to put sensors on the bearing pads so that they could detect exactly where the surface contour of the runner was deteriorating. Then, to keep the antenna running until a permanent correction could be made, they would insert thin metal strips called shims under the runner at the bad spots, to restore the critically important level surface of the runner (JPL 1974).

Phillips and his team believed that high spots had developed in the grout that supported the runner between its underlying sole plate and the top surface of the concrete pedestal. As each pad passed over these areas, they surmised, high-pressure oil escaped from beneath the pad, with a consequent reduction in the height of the supporting oil film. To correct the problem, they would move the antenna to a position where the pads were away from an identified high spot, lift the runner slightly, and insert shims between the runner and the sole plate to create a kind of ramp on either side of the bump. When the runner was lowered back into place, the pad would ride smoothly over the high spot without loss of oil pressure. It would be a temporary fix at best, but it would work for a while and allow time for further analysis to determine the real cause of the problem.

It is worth noting that the height of the high spots or bumps, and the thickness of the corrective shims, was no more than a few thousandths of an inch. Although the job sounded simple enough, it was in fact a significant task in precision mechanical engineering. Apart from the inherent difficulty of the task, it was difficult to get access to the antenna to carry out the work. Potter had the station running full time by then, and maintenance work such as this, which took the antenna out of service for hours or days, had to be carefully coordinated with all the other users demanding time on the antenna. It was difficult to persuade high-level spacecraft managers to give up tracking time on the antenna to allow time for what they perceived to be lower-level antenna maintenance. So resistant were they that it took an appeal to top-level management at NASA to resolve the near impasse. Edmond C. Buckley justified the work in a conciliatory letter to the deputy administrator, Seamans, and directed JPL to proceed (Buckley 1967).

Eventually the critical antenna maintenance work was scheduled for a few hours at a time on successive nights in March and April. Under the watchful eye of Horace Phillips and his team, mechanics carefully jacked up the runner, inserted the appropriate shims, and returned the antenna to service. Each time, the film height improved. It appeared that the problem had been solved.

JPL's *Mariner 5* spacecraft began its journey from Cape Canaveral to Venus in June 1967. The early stages of its journey were supported by the 85-foot tracking stations, but as it neared the planet, the full power of the 210-foot antenna, with six times the capability of the smaller antennas, was brought into play. During the Venus flyby on October 19, the spacecraft transmitted to Earth a wealth of new science data on the planet's surface temperature, atmosphere, and magnetic field via the powerful new antenna's downlink. The quality of the data and the speed with which it was received at Goldstone were considered remarkable accomplishments and a full vindication of NASA's support of the project.

A Very Good Antenna

Ever since he came to JPL in 1963 to work in the Antennas Group under Phil Potter, microwave engineer Dan Bathker had been associated with the microwave design and performance of the new 210-foot antenna. Although the antenna was not built yet, Bathker and his colleagues were busy designing microwave feed systems for it. Their initial designs were of course intended for use at S-band, the basic operating frequency for which the antenna was designed. When the antenna first went into service, a feed of this type was used to measure its radio frequency performance. The results were very good—so good, in fact, that they exceeded the specified design goal by a substantial margin (JPL 1974). In the minds of Bathker and his colleagues this immediately raised a question: If the antenna worked so well at S-band, how well would it work at X-band? In February 1968 Bathker got his chance to find out.

Although the advantages of operating a deep space link at X-band rather than the lower frequency corresponding to S-band were well known at the time, the benefits did not come free. There were very significant technical problems associated with designing functional X-band components, and until these were overcome, the operational network remained firmly settled on S-band as its operating frequency (Mudgway 2001). In the DSN's research and development facility at Goldstone, however, X-band was very much an active program. There, all kinds of X-band microwave devices and components were being designed, developed, and tested for possible use if and when the network moved to X-band. Among these components was a well-calibrated X-band feed cone system that, with minor modifications, would fit the adapter on the 210-foot antenna. When time became available, the change was quickly made and Bathker and his colleagues began to gather X-band performance data on the Big Dish. They were given seven days of antenna time, and they formed three teams to work round the clock. Dan Bathker remembered bringing several hundred strip-chart recordings back to JPL for analysis (Bathker 1992).

While the primary purpose of the X-band test was to evaluate antenna performance for possible future use, the shorter RF wavelength also provided a powerful medium for determining surface tolerances and other structurally related effects. The measurements confirmed the theoretical predictions of performance within the limits of error including the expected factor of almost ten times the gain of the antenna at S-band. When the analysis was finished, the final report concluded that "a significant ground capability for weather-dependent space communications or radio/radar astronomy is available at X-band using the 64-meter diameter antenna" (Bathker 1969).

Later, when asked for his opinion of the antenna, Bathker was reported to have said in his typical understated style, "You know, that really is a very good antenna." And who would have known better than he?[1] But as soon as Bathker's time was up, his X-band cone was replaced and the antenna resumed regular operation on S-band.

The year was to become a busy one. Early in 1968, Potter moved the entire S-band installation from the control room in the pedestal base, where he had run operations since the station first came into service, to a large new facility built specially for the purpose adjacent to the antenna site. At the same time, a significant amount of new equipment was added to handle the increased data-processing requirements of the two forthcoming Mariner missions to Mars. In addition to *Mariner 5* and several new Pioneer missions and the last of the lunar Surveyors, a manned flight mission appeared on the Mars station schedules for the first time. It was *Apollo 8,* the first manned lunar orbital mission. The six-day mission was part of the buildup of experience and technology that culminated in the actual Moon landing the following year. Although the Mars antenna played a backup role in these early manned missions, it was not until the later missions that its tremendous power would be amply demonstrated.

Get the Grout Out

Despite the efficacy of the initial "fix," the low oil-film height on the hydrostatic bearing continued to demand attention. Throughout 1967 and most of 1968, periodic shimming was required to maintain the desired film height, and it soon became apparent that something was dreadfully wrong with the grout. Special instrumentation to measure the size and exact location of the runner surface irregularities was installed and sample cores of the grout were analyzed. Horace Phillips and his engineers found that "the deterioration was the result of changes in the runner level due to oxidation of the sole plate at its interface with the grout." Simply put, there was rust between the steel sole plate and the grout that supported it on top of the concrete pedestal. Why? There were two factors. First, the special nonshrinking material that formed the grout contained iron filings and an oxidizing agent. Second, moisture from the atmosphere and from the frequent steam cleaning of the pedestal was penetrating the area between sole plate and grout. A complex chemical reaction resulted in the formation of iron oxide (rust) on the lower surface of the sole plate, which in turn deformed the runner surface (JPL 1974).

By the end of 1968, Merrick and Phillips decided to replace all of the original proprietary Embeco grout with a dry-pack sand-cement mixture. While this required much more care in preparation and curing, and more labor to hand-pack it properly, it contained no oxidizing material and was thus considered the best solution. Horace Phillips estimated that it would take six weeks. Despite screams of protest from the spacecraft managers, appropriate downtime was scheduled for January 1969.

The work sequence was carefully planned to minimize disruption of the critical Pioneer and Apollo tracking schedules, with the antenna remaining active, albeit with restricted azimuth motion, for the whole period of the work. That meant that the hydraulic oil could not be drained from the oil reservoir surrounding the runner and that the antenna would have to move over the new grout area within a few days of its placement. Using diamond-tipped bits operating simultaneously in the three spaces between the pads, workers drilled out the old grout and removed the rusted sole plates. New sole plates of improved design were placed in position and packed with the new grout. After allowing a minimum period for curing, technicians shimmed the runner to a level position and the workers moved on to the next section of the bearing. The work took eight weeks to complete, rather than the six originally planned. The antenna returned to service at the end of February 1969 (JPL 1974).

With *Apollo 11* scheduled to make the first Moon landing in June and two Mariner spacecraft due to arrive at Mars in July and August in addition to ongoing Pioneer support, Tom Potter's tracking schedule was completely filled. While the Apollo missions were comparatively short, a few days in duration, the length of the Mariner and Pioneer missions was measured in months and years. Both of these latter were high-profile events and required a great deal of equipment testing and operator staff training. In the case of Apollo, the Big Dish was linked via a complex microwave hookup to the Manned Spaceflight Network, whose 85-foot antenna was also situated in the Goldstone area some ten miles distant. In an elaborate data transfer arrangement, *Apollo 11* telemetry and television signals received by the 210-foot antenna were transferred to the MSFN for selective processing and public dissemination as determined by the Apollo mission controllers (Hartley 1970).

An important new feature of the later Mariner missions enabled imaging data to be returned from Mars in hours, rather than days as had previously been the case. This was achieved by making use of a thousandfold increase in data transmission rate—16,200 bits per second compared to 16 bits per second on previous planetary missions—which resulted for the most part from the improved receiving capability of the 210-foot antenna. However, this extra capability required not only the large antenna and its precision pointing system but also a significant increase in telemetry decoding, data processing, and recording equipment at the station. It was the station manager's job to see that all the equipment was functioning correctly, and that there was fully trained staff available to operate it as required by the tracking schedule.

To the immense satisfaction of its sponsors, designers, builders, operators, and not least its users, the great new antenna lived up to expectations and provided the critical tracking support to which it had been committed. It provided telemetry, and some imaging data, from *Apollo 11* during the critical landing and return-to-Earth parts of the mission, and also provided high-resolution photos from both Mariner spacecraft as they passed over the Martian surface at an altitude of 2,000 miles.

The Pioneer missions and a second Moon landing, *Apollo 12* in December, continued to engage the station through 1969. By then the Goldstone 210-foot antenna had become an established DSN capability, and back at JPL many new spacecraft managers were seeking, and getting, commitments from DSN managers to provide tracking support for their planetary missions. In no time at all, the network's single 210-foot (64-meter) antenna was oversubscribed.[2]

Tricone and Electronics

While Potter's technical staff at Goldstone dealt with the day-to-day problems associated with round-the-clock deep space operations—including the occasional low-film-height alerts that seemed to be starting up again—Stevens's engineers at JPL pressed forward with the development of innovative additions and improvements to the antenna that would enhance its S-band receiving capability even further. One of the notable improvements implemented in mid-1970 was known as the tricone.

Rather than a single feed horn, the tricone had three, each of a different type, mounted on the upper surface of the microwave feed structure. Each horn performed a special function and could be activated rapidly and simply by rotating the subreflector about its slightly offset axis to redirect the focused radio beam to the desired receiving horn. This scheme, which revolutionized the operational use of the antenna, was based on the concepts of Gerry Levy, Smoot Katow, and Charles Stelzried and is diagrammed in figure 7.1 (Bathker 1992; Stelzried, Levy, and Katow 1970).

From the very beginning, low-noise microwave amplifiers had been vital elements of the DSN's overall planetary communications system. To amplify the minute microwave signal collected by the receiving antenna, and to do so without degrading the signal by adding radio noise of its own making, was a task well beyond the capability of vacuum-tube amplifiers or the very best transistors. For that purpose, special solid-state devices called masers were required. Solid-state masers—their name derived from the descriptive phrase "*m*icrowave *a*mplification by *s*timulated *e*mission of *r*adiation—had been developed and demonstrated by groups at Bell Labs and at MIT as early as 1957.

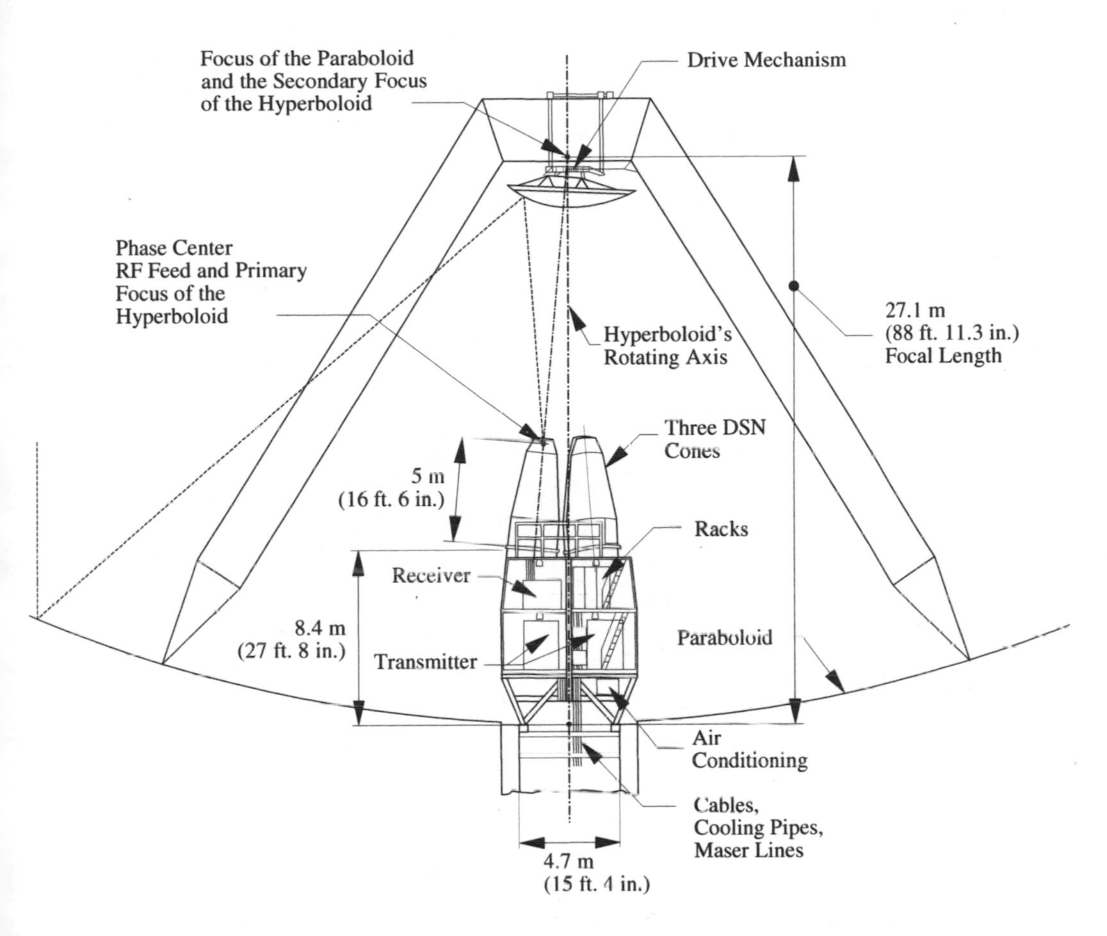

Figure 7.1. Configuration of 64-Meter Antenna Tricone and Subreflector. Although each of the three feed cones was displaced from the axis of the main paraboloidal reflector, the design of the hyperboloidal subreflector and its microwave geometry was such that rotation of the subreflector to illuminate any of the three feed cones introduced no boresight error into the pointing direction of the main beam. The tricone principle worked equally well when one of the feedcones was used for transmitting rather than receiving. Illustration by Fred D. McLaughlin, NASA/JPL.

Recognizing a possible application to the amplification of very weak signals from planetary spacecraft, engineers at JPL adapted and refined the basic principles to develop a maser specifically suited to DSN purposes. By 1962 a ruby-cavity maser operating at a frequency of 960 MHz had been installed in the 26-meter tracking station at Goldstone, where its superiority to the former cooled parametric amplifiers was demonstrated effectively on NASA/JPL's *Ranger 3* mission to the Moon in January of that year.

When the DSN moved its operating downlink frequency into the new Space Band (2.3 GHz, S-band) in 1963, the development group at JPL was ready with a new model of its maser, and by 1965 all three 26-meter antennas in the network were equipped with S-band masers.

These "traveling wave" masers and their closed-cycle refrigerator systems were commercial instruments that had been suitably modified for use in the field conditions typical of the DSN's remote 26-meter antennas. They were key elements of the system that captured the first TV pictures of the Martian surface at 8.3 bits per second from the *Mariner 4* spacecraft in 1965. With a tuning range of 2270–2300 MHz and a bandwidth of 17 MHz, these models produced a gain of 35 dB with a nominal system noise temperature of 55 kelvin.

By 1966, when the new 64-meter antenna was nearing completion, the DSN had accumulated considerable experience with operating such complex electronic devices in the field, although they were not troublefree. The maser and its associated electronic systems proved to have reliability problems, and the sophisticated nature of the equipment made field repairs very difficult and time-consuming. Nevertheless, it was decided to install this type of maser on the new antenna while JPL engineers embarked on a parallel effort to develop a more rugged and more stable model (Renzetti 1983).

The maser with its massive permanent magnet and closed-cycle refrigerator unit was mounted in the upper level of the tricone structure, close to the microwave feed horn. They were connected via a complex bundle of gas lines and control and monitoring cables to a large helium compressor and other vacuum and cryogenic support equipment on the ground. An intricate labyrinth of nitrogen-filled microwave waveguide structures and coaxial cables connected the maser to the feed horn and diplexer and delivered its amplified output to the down-converter circuits. The diplexer allowed for simultaneous reception of an S-band downlink and transmission of a high-power uplink at a slightly different but phase-coherent S-band frequency. The down-converter circuits used a phase-stable reference frequency from the station's frequency and timing system to generate a mixing signal for converting the received S-band signal down to an intermediate frequency of approximately 300 MHz. Low-loss coaxial cables delivered the 300-MHz signal from the tricone structure to the control room at ground level, far below. There racks of detectors, demodulators, decoders, data processors, and digital recorders and computers performed the intricate sequence of electronic functions required to extract and condition the science and engineering data for transmission to JPL.

In addition to the operational S-band maser, the tricone also housed an experimental X-band maser with all of its ancillary and support equipment. It too required waveguide and coaxial circuits and components to deliver its output to the data processing units in the control room far below. However, it was intended for receiving only and did not, at that time, involve a diplexer and an X-band transmitter. This maser had recently been developed by JPL based on its experience with earlier X-band models. It operated over the range 8370–8520 MHz, with 45–30 dB gain, a maser noise temperature of 7–13 K, and a 15-MHz bandwidth. Its installation on the new antenna would enable engineers to evaluate the performance of the new antenna at X-band and provide a field-testing environment for further improvements in what would eventually become a fully operational model for the network (Renzetti 1983).

The feed cone structure also accommodated several high-power microwave transmitters that formed the DSN's Earth-to-spacecraft communications link. This uplink consisted of a coherent S-band carrier signal that carried the spacecraft command and control data, and the ranging and Doppler reference frequency that was used for spacecraft navigation. Looking to the future in its design, the 64-meter antenna had made provision for supplying high-voltage DC power (70,000 V) and a large volume of deionized cooling water (800 gallons per minute) to the center level of the tricone structure, which was dedicated to the high-power transmitters and their supporting facilities. The primary power generators, transformer/rectifiers, and filter/choke components were located at ground level, as were the coolant supply, filters, pumps, and control components and the cooling towers. These facilities were part of the Rohr contract and were installed and available for use by the time the reflector surface was complete.

Initially, one standard DSN S-band transmitter consisting of a single Varian klystron tube with 10-kW RF output was installed in the transmitter module. It used power and coolant facilities already in place, and mated with the gas-pressurized waveguide runs to the diplexer. While it was sufficient for the immediate purpose of demonstrating the capability of the new antenna, it would soon be complemented by a second 10-kW unit, together with a 100-kW version that had been under joint development by Varian and JPL for some time. Later a 400-kW S-band version would be added to the ensemble for use in the event of a spacecraft emergency that might require the transmission of recovery command instructions on a superhigh-power uplink.

The interior of the tricone where these heavy, bulky, and extremely fragile devices had to be installed was very cramped, and the access was located more than 50 meters (164 feet) above ground level with the antenna pointed in the vertical position, adding to the difficulties of initial installation and subsequent maintenance. In normal use, the tricone tipped with the antenna through its full range of vertical rotation, so all the electronic equipment it housed was securely fastened to the interior framework to ensure its correct operation in any orientation. Nevertheless, the continual rotational motion posed a special problem for the masers, where the smallest movement of the heavy permanent magnet relative to the amplifying element of the unit affected the gain of the device in a significant way. Keeping these mounting structures sufficiently rigid under such conditions ultimately required a redesign of the entire traveling-wave maser and magnet structure.

The tricone and its microwave electronics added considerable weight to the antenna structure and raised a question in the minds of its designers. Would the additional loading perturb the precision of the carefully aligned tipping structure? Inevitably, Casperson responded to that question with an anecdote. "Dan Bathker was always raving about how good that antenna was. When he asked me why it seemed better after the tricone was added, I told him that Bill Merrick had always wanted a multiple feed arrangement, and so we added twenty thousand pounds to the original feed-cone support design requirements. We designed the structure with that weight in place, even though we did not know how to build a multiple feed cone system at the time. That was an example of the brilliance of Merrick's ideas" (Casperson 2000).

The implementation of the tricone on the 64-meter antenna in February and March 1970 substantially increased the station's flexibility for planetary radar and spacecraft tracking operations. When a high-power X-band transmitter was installed in the tricone the following month, the station's capabilities were even further enhanced. The microwave configuration of the station could now be changed from receive-only to receive-and-transmit or radar transmit-and-receive, for example, simply by resetting a few control panel switches. The time that was formerly consumed in the laborious process of changing feed cones was now devoted to increased tracking time, and the pace of network operations increased accordingly.

In April, during its then-routine tracking support for the Apollo, Pioneer, and Mariner Mars missions, station controllers suddenly received an emergency request for immediate continuous support of *Apollo 13,* en route to the Moon for the third manned lunar landing attempt. The unthinkable had happened. An oxygen tank explosion had forced the shutdown of the fuel cell used for generating electrical power, and the service module had become untenable. The mission had been aborted, and the three astronauts were desperately trying to return to Earth using the lunar module as an emergency lifeboat. To conserve power, radio transmissions from the module were limited to its low-power transmitter. Voice communications that were tenuous at best when received on the 26-meter antennas put an already risky Earth return attempt in even greater jeopardy. The significantly better sensitivity of the 64-meter antenna provided the compensation needed to restore normal communications between Apollo mission controllers and the astronauts and contributed materially to the happy outcome of the mission.

The DSN received a commendation from the director of flight operations at the Houston Manned Spaceflight Center: "We at MSC wish to commend the entire network for superior performance in support of *Apollo 13*. In the midst of this most difficult and critical mission it was extremely reassuring to have a network with so few anomalies, and one that provided us with the urgently needed voice and data to bring the crew back safely. We thank you for your outstanding support" (Hartley 1970).

By the end of 1970, the Goldstone 64-meter antenna had assumed its final form—or rather the form in which it would constitute a major network resource for the next decade. It was in continuous service for uplink and downlink deep space communications at S-band, and had demonstrated an ability to operate at X-band, although the latter capability was still under development. Aided immeasurably by the flexibility of the tricone which allowed rapid reconfiguration of the feed cones, it was being used regularly for planetary radar experiments and testing of new microwave components that were being developed to further enhance its awesome radio frequency capabilities. A spacious new control room housed hundreds of racks of electronic data processing, recording, timing, and communications equipment that performed the complex functions required to navigate and communicate with an increasing number of planetary spacecraft.

The antenna became a formal part of the operational Deep Space Network on January 21, 1971, when responsibility for the 64-meter Goldstone antenna was transferred from Merrick, representing the engineering organization, to Richard K. Mallis, the operations manager of the network. Mallis now looked to his operations engineers to keep the antenna up and running.

Wells

In mid-1966, as Goldstone's Big Dish began the transition from engineering status to operational status at the completion of the construction contract, a new cast of characters emerged upon the scene at the Goldstone Mars site. One of the more colorful, and more significant for the future well-being of the Big Dish, was Dale Wells.

Although Wells was new to JPL, he was by no means new to the 64-meter antenna. He had been an engineering supervisor in Bob Hall's new Antenna Division when Rohr bid on the Advanced Antenna contract in 1962. Subsequently he participated in the engineering design work for the alidade structure, the azimuth hydrostatic bearing and radial bearing arrangements, and, most important, the pedestal. He monitored much of the fabrication of the alidade structure and oversaw the trial erection of the alidade at Paramount. By then he knew, better than almost anyone else, how the complex structure went together and why it had been designed that way. Then, before the big steel started to move to Goldstone, he left Rohr to seek his fortune in the industrial world of precision machine tool design (Wells 1999).

Two years later, after that venture failed to meet his expectations, Wells began work at JPL as an engineer in Don McClure's group. Eventually he would become responsible for the maintenance of the structural and electromechanical parts of the antenna, with the title of Cognizant Operations Engineer. Dale Wells would spend the next twenty-five years of his professional life maintaining Goldstone's Big Dish, and the two others like it in Spain and Australia, in fully operational status.

Wells took part in Horace Phillips's early struggle with the hydrostatic bearing in a seemingly endless effort to keep the Goldstone antenna operating continuously. In January 1971, when he assumed responsibility for maintenance of the antenna, he introduced a rigorous program of regular bearing maintenance to address the continuing problems with oil film height that plagued the Goldstone antenna. While the bearing was stable for a year or so after the major regrouting episode in 1969, by 1972 it had become obvious that the grout was deteriorating again. Emergency repairs were called for, as *Mariner 8* was in Mars orbit and antenna time was critical. This time it took Wells and a team of mechanics six days of intensive round-the-clock effort to repair the worst sections of the azimuth runner. They would have to finish the job later. The following year they were at it again, this time for seven weeks. Aggravating as it was, there appeared to be no alternative. By 1976, antenna time for regrouting of the hydrostatic bearing had become a permanent feature of the DSS-14 maintenance schedule. Still the situation continued to worsen. Confidence in the reliability of the Goldstone antenna diminished to the point where, during the most critical mission sequences, Wells was required to stand by at the station to deal with any film height problems that might suddenly occur (Wells 1999).

Despite continuous attention by Wells and his mechanics, it had become apparent by 1980 that the entire process was losing ground and that the only viable solution would be to completely refurbish both the hydrostatic bearing and the pedestal that supported it. This would have been a major task even without the alidade and antenna in place, but it would be immensely more difficult to accomplish with them installed. How this challenge was met in the years ahead is a topic for the following chapter.

More Is Better

As the Goldstone 64-meter antenna came into service and began to demonstrate its outstanding performance in the late 1960s, a More Is Better initiative grew up at JPL and at NASA headquarters. Articles in public and professional journals advocated an increased capability for the United States' program of unmanned space exploration. One approach suggested a large number of smaller antennas (Drake 1965), the other a small number of larger antennas (Rechtin 1967). The Rechtin argument prevailed, and within NASA the initiative for two additional 64-meter antennas to complete the global network began to gain momentum.

The push for the two new 64-meter antennas was a prime example of the synergistic environment that pervaded NASA in general and the Tracking and Data Acquisition offices in particular in those years. The successful Mars flyby missions and the successful demonstration of high-data-rate telemetry by *Mariner 5* had stimulated the planetary science community to propose an ambitious mission to search for evidence of life forms on the surface of Mars. Plans called for two dual spacecraft, each comprising an orbiter and a lander, to visit the planet in 1976, ideally in time to demonstrate a United States presence on Mars for the bicentennial Independence Day celebrations. The two missions were to be called *Viking 1* and *Viking 2*. They would of course require high-rate telemetry and the maximum Earth-to-Mars communications capability for continuous twenty-four-hours-a-day operations. Here was the perfect justification for the two new antennas. The high public interest in a search-for-life mission, together with the demonstrated viability of the new antenna as evidenced by the recent outstanding imaging data from Mars, virtually guaranteed congressional support for Buckley's funding request. In December 1968, NASA announced its intention to seek proposals for the construction of two new 210-foot (64-meter) antennas, one in Australia and the other in Spain (NASA 1968).

Four companies responded, and six months later NASA chose the Collins Radio Company of Dallas, Texas, to build the two new overseas antennas for an estimated total cost of $20 million. Both antennas were to be similar to the Goldstone model. One was to be erected on the site occupied by NASA's existing Deep Space Station near Canberra, the other at the existing NASA station near Madrid.

The NASA news release said: "The new antennas will be fabricated in the United States and shipped to their overseas locations. The 5000-ton reinforced concrete pedestals on which the antennas will be mounted will be constructed on site. Fabrication is expected to get under way later this year, with the systems becoming operational in 1973, according to JPL." The announcement concluded, "The 210-foot antennas will send commands to, and recover data from, the various NASA planetary missions in the 1970s and 1980s, including the two Viking spacecraft planned for orbiting and landing on Mars. The 210-foot antennas provide more than six times the performance of the 85-foot dishes presently used in the Network, and will make possible the acquisition of scientific data from as far out in space as the edge of the solar system" (NASA 1969).

It was a bold prediction for its day, but in time the antennas did indeed live up to those high expectations, as we shall see.

Tidbinbilla

Every Australian summer they come, as they have come for thousands of years, millions of them, migrating to the cooler upland of the Yarralumla Plains in southeastern Australia to escape the heat of their breeding grounds on the lower flatlands. There they rest among the rocks and caverns along the Molonglo River before returning to the flatlands in early autumn to repeat their unending cycle of reproduction. The Bogong moths, for that is what the Aboriginal people called them, had fat, hairy, dark brown bodies an inch or two long and a wingspan of about two inches. Despite their unpromising appearance, when roasted in hot sand or ashes they made a delectable and nutritious source of food for the Aboriginal people. Every summer this area became a gathering place for the Koori tribes where the people would collect the resting moths from the sheltered sides of rocks and crevices and hold intertribal meetings while they feasted on Nature's providence.

From time immemorial the area was known in the Aboriginal language as Ngan-Girra, the gathering place. When the first of the European stockmen, overseers with assigned convicts, arrived about 1820 to establish cattle and sheep stations, the Aboriginal name became corrupted to Canbury or Canberry. By 1836 a permanent settlement had arisen near the Tuggeranong and Molonglo Rivers and its name had evolved into Canberra.

Canberra slumbered in the shadow of the eucalyptus-covered Brindabella Ranges until the turn of the century, when these penal "colonies" of Britain's far-flung empire were granted independence and united as the Commonwealth of Australia. One of the provisions of the new Australian constitution required the establishment of a site for a national capital. Canberra, conveniently situated roughly halfway between the archrival cities of Sydney and Melbourne, was chosen. The surrounding area was designated the Australian Capital Territory (ACT) in 1909. In an international competition two years later, the design by Walter Burley Griffin of the University of Illinois was selected for the new capital. Development of Griffin's radical design was slow, and it was not until 1927 that the first Australian federal government was convened at the new Parliament House.

The dream of a national capital began to move rapidly to fruition after World War II. The planned city straddled the Molonglo River, and over the next twenty years bridges were built across the river and the architect's imaginary lake. The National Mint was constructed, the lake was filled and named Lake Burley Griffin, the National Library and Botanic Gardens were added, and the Australian National University and the vibrant Civic Center were all completed in accordance with the original master plan. Through the 1960s the Australian Public Service moved many of its major departments to Canberra, and the city became a civil service city. Expanding rapidly from 50,000 in 1960, its population doubled to 100,000 by 1967. While Canberra attracted attention within Australia as an important new center of government, education, and population growth, at that time it was also attracting attention from a completely different quarter and for a completely different reason.

In 1960 Australia and the United States entered into a ten-year agreement on support for NASA's expanding space program in Australia. The United States agreed to meet the major costs of the program, while Australia would contribute $140,000 per year for local support. Subsequently NASA established several tracking stations around the country, one of which was the 26-meter deep space station at Woomera, a desert missile firing range 300 miles north of Adelaide in the state of South Australia. As a key element of the early Deep Space Instrumentation Facility,[3] the Woomera station supported JPL's Ranger, Surveyor, Lunar Orbiter, and early Mariner planetary missions with considerable success. By 1962, however, the need for additional NASA stations in Australia, together with the logistical complexities of maintaining a tracking station in an area as remote as Woomera, led to a joint investigation of suitable locations to establish new NASA facilities in more accessible parts of the country. The first requirement was for a second deep space station, but it was expected that a second manned-flight station and special stations for Earth satellite tracking would follow.

The search for a site in southeastern Australia was carried out in September 1962. The general choice of this region was driven by the longitude requirements for deep space stations, while the choice within the region was dictated by considerations similar to those that obtained at Goldstone—electrical noise-free environment, reasonable access to a town or city that could provide acceptable living and educational facilities for the technical staff and their families, and other factors such as freedom from natural disturbances like earthquakes, hurricanes, and floods.

For twenty miles the Cotter Road ran southwest of Canberra, out through the Tuggeranong, across the Murrumbidgee River, and past Cuppacumbalong where a wild kangaroo was more than likely to jump across the road and, in the early morning, brilliant-hued parrots cavorted and quarreled in the dense eucalyptus bush. Finally, as the bush opened out, the road sneaked into the rolling, open sheep pastures of the Tidbinbilla Valley. In ancient times this was a sacred place for the Ngunwal tribe. Here the tribespeople performed their ceremonial rites of passage from boyhood to manhood. In the Aboriginal language it was known as the Jedbinbilla, meaning "where all boys become men." In modern times it was about to become a keystone of the space age. It would be known then as Tidbinbilla (Bugg 2001).

There the survey teams found the ideal spot. Shielded from city-generated electrical noise by the surrounding low hills, a reasonable drive from a thriving and attractive population center, and an altogether idyllic location in which to work, the Tidbinbilla Valley was chosen as the site for the new deep space antenna in October 1962. NASA had come to Canberra (JPL 1965c).

Since all private property in the ACT was leased from the government, it was a relatively simple matter to withdraw the several hundred acres of land required for the tracking station from the leaseholder and construct an access road to connect with the Cotter Road to Canberra. The construction contracts for the buildings, facilities, and foundation for a 26-meter antenna were placed with local contractors, and Blaw-Knox received an order from JPL for another overseas 26-meter antenna. Driven by the pressing need for another antenna at that longitude to support the approaching *Mariner 4* encounter with Mars in 1964, work at the Tidbinbilla site progressed very rapidly indeed. Just over one year after construction began, the new Deep Space Station 42 (DSS-42) took over the task of tracking support for *Mariner 4* from the Woomera station, which then was fully engaged with the last of the lunar Ranger missions.

Australian Robert A. Leslie, a senior public service employee and a highly qualified electrical engineer of wide experience in managing advanced technical projects, was appointed director at Tidbinbilla. In formally opening the deep space station on March 19, 1965, Prime Minister Robert Menzies called the recent achievements in space "one of the miracles of the twentieth century; he acknowledged the advantages to Australia, as well as the United States, in having the tracking stations there but most of all he thought that the cooperation gave expression to the great friendship between the two countries" (Leslie 1980).

The DSN then had two active 26-meter antennas in Australia, both of them fully engaged with JPL's lunar and planetary missions. In due course NASA constructed two more tracking stations in the general area of Tidbinbilla, one at Honeysuckle Creek for tracking the Apollo manned space missions, the other at

Orroral Valley for tracking NASA's Earth-orbiting satellite missions. It was a busy time for NASA in Canberra.

Under Leslie's direction, the Tidbinbilla station soon became an integral part of the network. The high quality of its technical and operations staff contributed greatly to overall performance of the network over the ensuing years. Following the *Mariner 4* mission to Mars, it was involved with the *Pioneer 6* solar probe, for which it carried out the "initial acquisition" of the downlink signal following the launch from Cape Canaveral. Its unique geographic location satisfied the complex conditions required to perform this most critical task, and its success with *Pioneer 6* immediately established its reputation as the network's initial acquisition station. The first report of "receiver in-lock," signifying that DSS-42 had found and acquired the downlink from a just-launched spacecraft, brought tremendous sighs of relief from spacecraft controllers at JPL for many years thereafter. The station went on to support the Surveyor lunar lander spacecraft and the later Mariner missions to Venus and Mars in 1967, 1969, and 1971.

In 1969 Leslie was transferred from Tidbinbilla to head the newly created American Projects Branch of the Department of Supply at its central office in Canberra. Tom Reid replaced him as station director, and his deputy, Frank Northey, was given responsibility for a second antenna then planned for construction at the Tidbinbilla site. It was to be 64 meters in diameter, similar to the Big Dish at Goldstone, and would be known as Deep Space Station 43.

The construction and erection of the Tidbinbilla antenna followed pretty much the same sequence as at Goldstone, except of course for deviations due to unexpected incidents, anomalies, and local conditions. In another example of Merrick's depredations upon industry for key personnel, he hired Bert Sweetser, an engineer who had been with the firm that did the architectural and engineering design for the Goldstone antenna, as his on-site resident engineer.

The Australians wasted no time in getting started. A month after the NASA announcement, their local contractors were placing concrete for the instrument tower footings. At the beginning of October 1970 the pedestal was complete and ready for installation of the azimuth hydrostatic bearing runner. The runner was grouted into place and leveled and the Collins ironworkers began erecting the alidade structure in February 1971. Almost exactly a year later, in January 1972, the new antenna performed its first elevation rotation. The antenna began tracking spacecraft in July 1972 as part of the contract acceptance test formalities. It took another month to close out the contract work, and by August 1972 the contractors had left the site. A large new control room was already available on-site, and the electronic equipment was installed directly into its permanent location. By the end of the year all of the S-band equipment had been installed and tested. On March 30, 1973, the new Australian 64-meter antenna was transferred from engineering status to full network operational status. Much to JPL's relief, there were no problems with the oil film height on the Tidbinbilla hydrostatic bearing as there had been at Goldstone. The photograph in figure 7.2 shows the completed 64-meter antenna at the Canberra Deep Space Communications Complex.

Robledo de Chavela

As the requirement for an additional antenna near each of the three principal longitudes became more evident toward the end of 1961, considerations of potential risk to future NASA facilities persuaded JPL and NASA that new antennas on the longitude of the existing station near Johannesburg should be placed somewhere in Europe, rather than in South Africa. The desire for an alternative site arose not from any dissatisfaction with the operation of the Johannesburg station, which indeed had an exemplary record of performance, but rather from a growing public uneasiness in the political relations between the U.S. and South African governments following the 1960 racial disturbances. Reflecting this uneasiness, NASA maintained a low profile about its activities in South Africa, unlike those in Australia, which had been well publicized. Nevertheless, it was clear that NASA intended to have a new station in operation at that longitude in case it should eventually wish to withdraw from the Johannesburg site.[4]

Some years earlier, in 1959, an American group acting under the auspices of the Advanced Research Projects Agency had visited Spain to seek suitable sites for future ARPA stations. Diplomatic agreements between the United States and Spain limited their survey to existing U.S. military bases, where the high level of electronic interference obviously rendered them unsuitable locations for sensitive receiving equipment. For this reason, South Africa was chosen as the site for the first deep space station at that longitude.

However, things had changed by 1962. NASA and ARPA had become independent agencies, with NASA devoted entirely to the civilian space program and ARPA to space-related military projects. The Spanish government's Instituto Nacional de Técnica Aeronautica (INTA) had participated in NASA's manned Apollo program by successfully operating a tracking station in the Canary Islands off the coast of Africa and had indicated its interest in further involvement with NASA. The stage was set for a new approach.

NASA headquarters made the first move. At the end of 1962, Edmond C. Buckley formally requested the State Department's approval for a NASA-sponsored visit to Spain to brief Spanish government officials on the requirements for a deep space tracking site and to investigate several areas of interest. Despite some misgivings about entering into discussions with the authoritarian Franco

Figure 7.2. DSN 64-Meter Antenna at Tidbinbilla near Canberra, Australia, 1972. The contrast in size of the new 64-meter deep space communications antenna and the older 26-meter antenna is clear in this side-by-side photo taken shortly after completion of the 64-meter antenna. Collins Radio Company, as a prime contractor to NASA, constructed the new antenna between November 1969 and July 1972. NASA/JPL.

government, State Department officials approved the request. Rechtin, representing JPL, and Gerald Truszynski, representing NASA's Office of Tracking and Data Acquisition, left for Spain in January 1963. They hoped to find a well-shielded site within an hour's drive of a major city and a ready source of engineering and technical staff to operate a future NASA tracking station (Tardani 1963).

Their expectations were soon fulfilled. Truszynski and Rechtin returned from their five-day visit confident that suitable sites for both a deep space and an Apollo tracking station existed within easy driving distance of Madrid, the capital city. They were particularly interested in an area near San Martín de Valdeiglesias, on a main highway thirty miles west of Madrid. Within a few months, technical teams from JPL and INTA had conducted detailed radio frequency interference surveys of all the possible sites within the area and made their recommendations. JPL and NASA eventually chose a site to the west of a high ridge that provided a natural barrier to shield the new antenna from electrical noise sources in the Madrid area. By mutual consent, INTA and NASA elected to call the chosen site Robledo de Chavela after the nearby pueblo of that name (Tardani 1963).

Madrid was little more than a small village in 1492 when the Spanish monarchs, Ferdinand and Isabella, financed the voyage of the Italian mariner Christopher Columbus, which resulted in the discovery of the New World. For its time, it was a national enterprise of scope, magnitude, and ambition that paralleled the missions of space exploration of modern times. By the middle of the following century Spain had become a superpower of the age. She was at the height of her political and economic power. In symbolic gestures of this wealth and influence, magnificent edifices and monuments were built throughout the country. When the reigning monarch, Philip II, selected this plateau in the geographical center of the country for the site of a new royal residence, Madrid began to develop in size and importance to eventually become the capital and largest city in Spain.

Named with the Spanish word for an oak grove, Robledo de Chavela was located at an altitude of about 2,500 feet in a small valley along the slopes of the Sierra de Guadarrama, forty miles northwest of Madrid. For centuries, kings and Madrileños had retreated to Robledo and similar pueblos in these mountains to escape the summer heat of the city. In the mid-sixteenth century, King Philip II elected to build a magnificent palace/monastery near there as a symbol of the power and might of medieval Spain. For twenty years the construction of San Lorenzo de El Escorial dominated the Spanish economy. Prominently situated on the flanks of the Guadarrama, the gigantic monument included twelve acres of buildings, connected by more than a hundred miles of corridors. During his lifetime Philip directed the Inquisition from El Escorial, and when he died in 1598 it became his mausoleum. In modern times, thousands of visitors tour El Escorial each year to wonder at the magnificence of this towering monument to the extravagances of the sixteenth-century Spain (Chamarro 2001).

It took a further year for the necessary international agreements to be completed, and for NASA and INTA to agree on exactly how the station would be manned and operated. Unlike at the other sites, there was the added difficulty in Spain of finding enough qualified technical personnel with fluency in English, a necessity for worldwide DSN operations. Eventually it was agreed that JPL personnel would assist INTA personnel in the initial operation of the station and would gradually phase out as the Spanish staff gained proficiency and experience.

Spanish contractors completed the site preparation and construction of the foundation for the antenna by the end of 1964. While Blaw-Knox erected yet another 26-meter antenna, Spanish stonemasons built a splendid granite building to house the main control room and supporting facilities, and a joint JPL-INTA team installed the electronics. Following a system checkout in May, the new station under its first JPL manager, Don Meyer, supported by Spanish engineers and technicians, became operational in time to lend its support to the *Mariner 4* flyby encounter with Mars in July 1965 (JPL 1965b).

Subsequently Rechtin chose two more sites in the same general area, one for a second deep space station and one for an Apollo station. The stations that NASA erected at these sites were known as Cebreros and Fresnedillas after small villages nearby. Initially the stations were operated by a combined Spanish and American staff in accordance with the intergovernmental agreement. Later, when the Spanish technical staff had acquired the necessary experience both on the job and in training courses at JPL, INTA took over the management of the Cebreros station in June 1969 and Robledo de Chavela in June 1970.

It was therefore with considerable pride in its accomplishment and confidence in its technical resources that INTA prepared to accept the significant additional responsibility associated with NASA's announced intention to construct a 64-meter antenna at the Robledo site.

The construction work in Spain overlapped the work in Australia by deliberate intent. On June 18, 1970, representatives from NASA, JPL, Collins Radio, and INTA met in Madrid to discuss plans, schedules, and initiation of the project. Merrick had chosen an engineer named Ken Bartos to oversee the work in Spain on behalf of JPL. Bartos would be the counterpart of Sweetser in Canberra and McClure at Goldstone. The construction work began slowly, owing to adverse foundation conditions that required a considerable amount of blasting to excavate the solid granite for the pedestal and instrument tower foundations. But once it started, the work proceeded through construction of the pedestal, setting of the hydrostatic bearing, and erection of the alidade without major interruption.

Later, however, aggravating incidents began to impede completion of the work. One night in May, unusually high winds at the site damaged a number of the delicate aluminum ribs that were stacked upright on the ground prior to assembly on the reflector structure. Many of them required rework. Despite such incidental delays, Bartos was able to renegotiate the remaining work schedules with the contractor to bring the important azimuth and elevation rotation milestones in on time in February and September 1972, respectively. Back at JPL, Merrick was well satisfied with Bartos's regular reports and believed that the Madrid project was moving rapidly toward a successful conclusion. No one could have foreseen what happened next.

During a routine azimuth rotation test, a foreign object became caught between the runner and a film height sensor on one of the pads. Before the antenna could be brought to a stop, a 20-foot-long section of the runner had been gouged and the matching pad surface had been damaged across its full length. Although the cause was different, this accident was very similar to the one at Goldstone during its construction, and similar techniques were employed to make the repairs. The corner weldment of the alidade was jacked up, the damaged pad and runner section were removed, and in a carefully temperature-controlled welding process they were resurfaced and reground to the original specifications for flatness. It was a splendid effort on the part of the Spanish technicians who performed the delicate and exacting work. Three weeks later the antenna was returned to service and the acceptance testing process resumed (Bartos 1975).

JPL accepted the new antenna from the contractor one month later, on January 2, 1973. Although the antenna was immediately put into service on live spacecraft tracking, as with the Australian antenna, considerable rework remained to bring the antenna up to full operational status. This was accomplished over the ensuing months, and by midyear the Spanish 64-meter antenna was transferred to fully-operational status. The Madrid Deep Space Communications Complex, as it appeared shortly after the completion of the 64-meter antenna, is shown in the photograph (figure 7.3).

Finally, in addition to its two existing subnetworks of 26-meter antennas, the DSN had a subnetwork of three 64-meter antennas spanning the globe. The Deep Space Stations were designated DSS-14 (Goldstone), DSS-43 (Canberra), and DSS-63 (Madrid).[5] They arrived on the scene not a moment too soon, for their superb communications capabilities were in great demand.

Before a Decade Was Out

In the four-year period from late 1969, when construction began on DSS-43, until late 1973, when work on DSS-63 was completed, a great deal of attention had been focused on the expeditious accomplishment of those two tasks. Important as they were, however, they were but a sidebar to the main stream of network activity, the tracking of NASA's ever-increasing array of planetary spacecraft. As each new antenna became operable, it was immediately put into service to relieve the gross oversubscription of the network's existing resources.

The growth in activity was largely driven by the fact that most planetary spacecraft did not simply "go away" after accomplishing their principal scientific objectives, such as the flyby of a planet or a prescribed number of orbits, but continued to operate perfectly well and maintain their ability to observe and

Figure 7.3. DSN 64-Meter Antenna at Robledo de Chavela near Madrid, Spain, 1973. The relative sizes of the 64-meter and 26-meter deep space communications antennas are apparent. The larger antenna has more than six times the radio receiving capability of the smaller antenna. Collins Radio Company, as a prime contractor to NASA, constructed the new antenna between June 1970 and January 1973. Prince (later King) Juan Carlos of Spain attended the formal dedication of the new antenna in May 1974. NASA/JPL.

collect scientific data. Such conditions proved irresistible to the science community, who pressured NASA to support "extended missions," albeit with lower priority than "prime missions," to complement the data already gathered. While the extra data came at little or no cost to the spacecraft managers, it posed serious problems for the DSN in accommodating the additional tracking time that the extended missions required.

Overlaying the missions already in extended status, the *Mariner 9* spacecraft in 1971 became the first to orbit another planet, in this case Mars. During almost a year in orbit, *Mariner 9* used the DSS-14 antenna and the high-rate telemetry system demonstrated in 1969 by *Mariner 7* to return detailed photographs of the two Martian moons, Phobos and Deimos, and high-resolution maps of the entire surface of the planet. Launched two years later in 1973, *Mariner 10* became the first spacecraft to visit two planets. In 1974 it used the gravity of Venus to execute an encounter with Mercury and returned the first ultraviolet photos of Venus and close-up images of Mercury's surface features. The new 64-meter antennas combined with the shorter range of Venus to permit an in-flight demonstration of X-band downlink performance at telemetry data rates up to 116,000 bits per second, nearly ten times greater than previous rates. The Apollo flights continued to demand attention from DSS-14 until they ended with *Apollo 17* in December 1972. Also during this period, Pioneer missions continued to make space history with the flybys of Jupiter, the first by *Pioneer 10* in 1973, the second by *Pioneer 11* in 1974. *Pioneer 11* went on to pass Saturn in September 1979. Both of these missions depended on the 64-meter antennas for their unprecedented long-distance links with Earth. As extended missions, both spacecraft would make use of improved DSN capabilities to compensate for increasing distance for many years to come.

While the completion of the 64-meter subnet in 1973 marked the beginning of a new era in the communications capabilities of the DSN, the completion of the Mariner missions in 1975 marked the end of the older generation of spacecraft. The spacecraft that replaced them were specifically designed to match the DSN's enhanced capabilities. They were bigger, heavier, more capable, more complex, ran at higher telemetry data rates, depended on the 64-meter antennas for tracking and data acquisition, and were, of course, more costly.[6]

The Viking spacecraft were the first of the big new generation to appear on the DSN tracking schedules. Comprising two orbiters circling Mars simultaneously with two active landers on the surface, Viking dominated the attention of the network from well before the first launch in mid-1975 until about 1980, when the last of the four spacecraft ceased operating. All four successfully completed their short (three-month) primary missions and continued operating for several years to return even more spectacular data as extended missions. During their lifetimes the orbiters returned 52,000 images covering 97 percent of the Martian surface, while the landers, operating from widely separated locations on the Plains of Chryse and Utopia, returned 4,500 close-up images of the Martian landscape and provided more than three million weather reports.

Just two years later, in 1977, NASA launched the second set of its new-generation planetary spacecraft. Known as *Voyager 1* and *Voyager 2*, their original objective was to investigate the Jupiter and Saturn planetary systems. Ultimately their itinerary also included Uranus and Neptune. These missions really pushed the DSN space communications capability to the outer limits. Because of the enormous distances involved and the commensurately long flight times, the various planetary encounters did not occur until long after Viking operations had ended. Although the actual encounter sequences were quite short, a few days in duration, the pre- and postencounter events extended for many weeks or months in some cases, and they too required the undivided attention of the large antennas. Together with long periods of uninterrupted downlink tracking required to generate navigational data of sufficient precision to ensure the desired encounter geometry, the Voyager mission requirements for tracking and data acquisition on the DSN were awesome indeed. Voyager became a permanent fixture on DSN tracking schedules for the next twenty years.

Both *Voyager 1* and *Voyager 2* reached Jupiter in 1979. Using a complex gravity-assist technique, *Voyager 1* was first to Saturn in 1980, followed a few months later in 1981 by *Voyager 2*. *Voyager 1* continued on a trajectory that would eventually take it out of the solar system into interstellar space. Further application of the gravity-assist maneuver redirected *Voyager 2* to close encounters with Uranus in 1986 and Neptune in 1989. Both spacecraft investigated the Jupiter and Saturn planetary systems and returned thousands of startling and beautiful images of these bodies and their satellites as they flew by. *Voyager 2* extended these investigations to Neptune and Uranus before it too began its long exit from our solar system.

The end of the 1970s marked a turning point in the race with the Soviets for preeminence in both lunar and planetary exploration. Preceded by the successful lunar orbiters and the Surveyor unmanned landings on the moon in 1966–68, NASA's Apollo manned landings had captured world attention from 1969 through 1972. At the same time, the Soviet lunar program had experienced mixed success with its lunar orbiters and landers. It had, however, succeeded in returning a sample of lunar soil to Earth.

In that era NASA's planetary spacecraft were reaching ever further into deep space and returning dazzling images and exciting new science data about other planets in the solar system. Mercury had been visited by *Mariner 10*, Venus had been probed by Pioneers and Mariners, Mars had been visited by Mariners and explored by Viking orbiters and landers, and *Pioneer 10* and *11* had flown by Jupiter and Saturn to pave the way for the two Voyager spacecraft that followed a few years later.

The awesome power of the Deep Space Network not only enabled missions such as these to be navigated to distant planets with great accuracy and to return their scientific observations to Earth with great rapidity, it also attracted the attention of the science community as an advanced instrument for research in the field of radio astronomy.

Meanwhile the Soviets' planetary program had remained focused on Venus with some success but, despite repeated attempts, had not been successful with its Mars missions (Perminov 1999).

So it was in this context that the decade of the 1980s began with the perception that the preeminence in space that the United States had so desperately sought two decades earlier had been amply demonstrated. This change in attitude translated into a different set of priorities for the Deep Space Network, and the motivation for a "bigger and better" network came, not from an international space race, but from the spacecraft themselves and from the challenges of the deep space environment in which they moved.

Bigger Is Better

Changing Times

The composition, capabilities, and challenges of the Deep Space Network changed significantly in the years following the completion of its subnetwork of three 64-meter antennas in 1973. No longer constrained by the tracking capabilities of the older 26-meter antennas, NASA scientists proposed a plethora of ambitious missions to explore Venus, Mars, Jupiter, Saturn, and even Uranus and Neptune, all of them dependent upon the awesome power of the DSN's new 64-meter antennas for their uplink and downlink communications with Earth. To complement the signal-gathering capability of the 64-meter antennas, engineers at JPL designed new microwave, telemetry, command, and navigation systems that would meet the ever-increasing demands of the proposed new missions. These designs translated into huge quantities of sophisticated hardware and software that were to be installed at the three sites around the world. Detailed planning of the complex logistics, installation, and testing sequences ensured that the additions, upgrades, and reconfigurations were accomplished without interruption to ongoing in-flight operations.

In the late 1970s the spacecraft design problems that resulted in the failure of many of JPL's early missions had been overcome, and the life expectancy of planetary spacecraft was determined more by the supply of attitude control gas than by mystifying system failures. Scientists convinced NASA that, even though the prime mission objectives had been accomplished, surviving spacecraft could still return valuable scientific data, and new "extended missions" were designed for such spacecraft. Each extended mission carried with it an additional set of calls on the DSN for tracking and data acquisition support. Thus, in time, the DSN accumulated responsibility for tracking more and more spacecraft, not always sequentially but more often than not simultaneously. Multimission capability—the ability to process telemetry and command and navigation data from several antennas, each tracking a different spacecraft, simultaneously at a single site—became a prime engineering objective in the network.

This situation forced the DSN to develop more intelligent and, regrettably, more restrictive practices for allocating its antenna tracking time. The addition of more antennas was the obvious answer, and the completion of the 64-meter network in 1973 helped alleviate the situation to some degree. It was not long, however, before it too became oversubscribed with missions that were beyond the capability of the 26-meter stations, and the DSN began to search for ways to ease the load on the 64-meter network. Eventually an engineering solution to this apparently intractable problem was found in a plan to modify the existing 26-meter antennas. By increasing their diameters to 34 meters, their performance was improved to the point where these formerly 26-meter antennas could carry a great deal of the tracking load previously assigned to the 64-meter antennas. Over a period of eighteen months from November 1978 through April 1980, one 26-meter antenna at each site was converted to 34-meter diameter and, together with improvements to its servo mechanical and microwave systems, returned to service as an element of the new 34-meter network.

The technical changes that the DSN saw in the 1970s were accompanied by many changes in its organizational structure and in the engineering personnel that managed it. Eberhardt Rechtin, whose purposeful vision was instrumental in securing full NASA support for the Advanced Antenna System, left JPL in 1967 for a position with a defense research agency, and Robertson Stevens as new chief of the Telecommunications Division became responsible for all engineering development related to the DSN. Bill Merrick, promoted to management level in 1965 during construction of the Goldstone antenna, received NASA's Exceptional Service Medal for his contribution to the then-completed 64-meter antenna project. He spent the rest of his career at JPL working in other areas. The position of manager of the DSN Antenna Engineering Section was taken over by his deputy Floyd Stoller, a highly qualified engineer of long experience with the DSN, while Dr. Peter T. Lyman, a strong, talented, and experienced technical manager who had long been associated with JPL's planetary spacecraft programs, became director of JPL's Tracking and Data Acquisition Office. As director, Lyman relied upon Bob Stevens to advise him on engineering matters and upon program manager Carl Johnson to advise him on budgetary and NASA funding matters. It was a very powerful organization within JPL and exercised considerable influence at NASA headquarters.

An Array of Antennas

The DSN had long been interested in enhancing the quality of the signal from a single antenna by combining it with the signal from a second antenna. This technique, known as "arraying," depends for its success upon the fact that the radio noise that inevitably degrades, or corrupts, the radio signal from a distant spacecraft is incoherent, or random, while the phase of the signal is coherent, or constant. When the outputs of two separated antennas, each containing the signal plus a proportion of noise, are properly combined, the incoherent noise components tend to cancel, while the coherent signal components simply add. The net effect is to enhance the ratio of signal power to noise power produced by the array, relative to that of either antenna alone. In effect, the radio flux collecting area of the individual antennas has been synthesized into that of a single antenna of size approximately equal to the sum of the two. Although the concept is simple, it is very complex indeed to add the outputs of the individual antennas with sufficient precision to make the scheme work. There are two principal ways of carrying out this critical "adding function."

In the first method, called carrier-arraying, the radio-frequency carrier signals at S-band or X-band are summed by elements of the microwave system and passed to a phase-tracking receiver for processing. The enhanced signal-to-noise ratio at the receiver aids the carrier-tracking process and improves the subsequent data processing functions. Overall this reduces the number of erroneous bits, or bit error rate, in the output data stream—that is, it improves the quality of the resultant data products. In practice, however, it is extremely difficult to maintain phase coherence between the two carrier signals at the input to the summing device when the radio path length between the two widely separated antennas can be as much as hundreds of meters—or, in the case of Parkes, hundreds of kilometers. Variations in radio path length can quickly degrade the entire phase-matching process enough to offset the carrier-arraying advantage.

These difficulties are bypassed in the second method, known as baseband or time-matching arraying. In this scheme the widely separated antennas with their S-band or X-band receivers function in the normal way to produce a data-modulated signal at the subcarrier or baseband frequency, generally on the order of a few hundred kilohertz. The independent baseband signals are then conveyed by landline, satellite, or microwave link to a central location where the digital combining process is to be carried out. At baseband frequencies, variations in time delay from the Earth communications links have much less effect on the ultimate combining process. The real-time combiner introduces a fixed delay to account for these Earth links plus a continuously variable time delay to compensate for the change in signal arrival time at the different antennas due to Earth rotation during the spacecraft tracking period. Once they are properly time-adjusted, the two baseband signals are summed digitally by a mathematically weighted function based upon their signal-to-noise ratios. Finally, the enhanced baseband signal is presented for subcarrier demodulation prior to decoding and subsequent steps in the data handling process.

Because of its obvious advantage for widely separated antennas, the DSN adopted the baseband arraying technique for the network, subject to a successful field demonstration during the Voyager encounters with Jupiter (Stevens 1983).

By the time the Voyager spacecraft reached Jupiter in 1979, DSN engineers at Goldstone were ready to demonstrate the technical feasibility of arraying a smaller 26-meter antenna with a large 64-meter antenna to enhance the stand-alone performance of the latter by about 1 dB, which worked out to an improvement of 25 percent (or, in antenna parlance, a factor of 1.25). Amid a wealth of other science data, this arrangement produced stunning images of Jupiter like that shown in figure 8.1. The photograph is one of thousands transmitted by the Voyager spacecraft from Jupiter to Earth via the 64-meter antennas of the Deep Space Network.

Confident in the viability of the arraying technique, the DSN then implemented a multiple antenna arraying capability at all three complexes around the network and used it with similar success to capture the science data from the Voyager encounters with Saturn in 1980–81. Despite the enormous increase in transmission distance,[1] the science data and images received from the Voyager spacecraft were of extremely high quality. If the DSN engineers were excited, the Voyager scientists were ecstatic—and with some justification, as the breathtaking image of Saturn's rings in figure 8.2 demonstrates.

Bruce Murray, who was JPL's director at the time, later characterized the events surrounding these encounters as "A Sublime Rendezvous with Saturn" (Murray 1989). Each of the two Voyagers returned about 15,000 television images of Saturn, plus an avalanche of other new science data about the Saturnian system. In the great excitement over the new science and beautiful pictures of our distant neighbor in the solar system, it was easy to overlook the exquisite intricacy of the radio links that brought them to Earth. Traveling at the speed of light, over a radio path almost 1.5 billion kilometers in length, the data took

Figure 8.1. *Voyager 1* Close-up of Jupiter. One of the most spectacular planetary photographs ever taken, this was obtained on February 13, 1979, as *Voyager 1* approached Jupiter at a range of 12.5 million miles. Two of the four Galilean satellites, Io on the left and Europa on the right, are passing in front of the planet. NASA/JPL.

Figure 8.2. Rings of Saturn. The *Voyager 2* spacecraft is looking back at the dark southern side of the rings in this picture taken August 29, 1981, from a distance of 2.1 million miles. The bright disk of the planet is visible through the tenuous C ring while the denser rings cast a dark shadow upon the planet's bland surface. NASA/JPL.

close to an hour and a half to reach Earth. At the end of its journey, the radio signal that carried the precious imaging and science data was unimaginably small.[2] It was altogether an amazing achievement for a radio link that began at the planet Saturn with a spacecraft transmitter having the power of a 25-watt refrigerator light.

But, for the DSN, the saga of Voyager was not to end there. It was, in fact, only just beginning.

Influenced almost entirely by the gravitational effects of its encounter with Saturn, one of the spacecraft, *Voyager 1,* altered course slightly and began a decades-long journey that would eventually take it clear out of the solar system and into interstellar space. Subjected to the same pervasive force, its sister *Voyager 2* made a more abrupt turn around Saturn and began traveling along a new trajectory that would take it to a close encounter with Uranus, twice as far again from Earth, in 1986.

The possible extension of the *Voyager 2* mission to an encounter with Uranus in 1986 immediately created an apparently intractable problem for the DSN. Uranus was 20 AU away, twice as far as Saturn, and there was no obvious way of providing the 6 dB (factor of 4) improvement in performance required to compensate for the doubled transmission distance. All the known methods of increasing the performance of existing facilities, such as arraying, had been used up. And the paucity of new planetary missions in that period of NASA history precluded any new antenna construction. The solution, whatever it was, would have to come from existing resources. It was a challenge on which JPL engineering expertise had thrived in the past, and a two-part solution was soon found.

First, the spacecraft engineers invoked a new data processing technique called data compression to effectively reduce the number of redundant bits that were normally embedded in the data stream from their spacecraft.[3] This improved efficiency in the telemetry data stream and translated into a substantial reduction in the strength of the downlink signal that was required to carry the telemetry-borne science and image data. Alternatively, the same quality data could now be transmitted from a greater distance without additional degradation. Although the data compression technique helped enormously, it still did not make up the full 6 dB that was needed. There remained a shortfall of 2 dB (a factor of 1.6), which the DSN engineers had to find. And find it they did—in Australia.

As was mentioned earlier, there had long been a cordial working relationship between NASA's DSN facilities at Canberra and the Australian Radio Astronomy Observatory at Parkes. By 1982 both facilities had 64-meter antennas, albeit of quite different design, and the two collaborated and shared antenna time to their mutual advantage. Now DSN engineers proposed to carry this collaboration a step further. They would augment the DSN 64-meter antenna performance with the Parkes 64-meter performance. Calculations showed that arraying these two 64-meter antennas with two 34-meter high-efficiency antennas then being built at Canberra could provide the receiving capability necessary to give a satisfactory science data yield from *Voyager 2* at Uranus distance. It seemed that data compression together with DSN-Parkes arraying could do the job, and so it was decided.[4]

Three years and a lot of hard work later, the Parkes-Canberra Telemetry Array was ready to capture the science and imaging data from the most distant planet yet observed close-up by the prying eyes of an inquisitive spacecraft. The photographs of Uranus revealed the shallow hydrogen-helium atmosphere to be relatively inactive, with no visible hurricanelike storms, and the weak ring system displayed no particularly remarkable features. However, *Voyager 2*'s closest approach to Uranus on January 24, 1986, also brought the spacecraft near one of the planet's little moons, named Miranda, and without great expectations for what they might reveal, a series of close-ups were taken. When the DSN delivered the images to Pasadena, scientists were astounded. Miranda, they said, was the scientific equivalent of striking paydirt. A typical image of Miranda, received by the Parkes-Canberra array from the spacecraft *Voyager 2* more than 20 AU from Earth, is shown in figure 8.3.

Satisfied that the immediate problem of providing tracking and data acquisition support for the *Voyager 2* encounter with Uranus had been solved, Stevens and his engineers began considering what might be done to extend the reach of the network even further. Spacecraft orbit designers had already shown that they could, and intended to, navigate *Voyager 2* to the next logical waypoint on its grand tour, Neptune. At its encounter with Neptune, a chasm of nearly 3 billion miles of deep space would separate the spacecraft from the tracking antennas on Earth. For the spacecraft engineers, extending the communications distance yet again, to 30 AU, would require careful attention to maintain the craft in working condition while it moved inexorably along its new trajectory toward the encounter point. But for the DSN it posed a quite different problem.

Stevens stated the dilemma then facing the DSN: "The Voyager signals from Neptune will be 3.5 dB [more than a factor of 2] weaker than from Uranus. To achieve comparable data yield, that deficit must be made up. There is no further potential improvement available from the spacecraft; in fact, with the aging spacecraft, the opposite is true. Increasing the ground aperture is the only evident way to reduce the deficit" (Stevens 1984). Alluding to the hiatus in new NASA planetary mission initiatives, Stevens continued, "Except for the Voyager Neptune encounter [in 1989], the DSN does not now perceive any spacecraft mission requirements for truly exceptional telemetry support throughout the

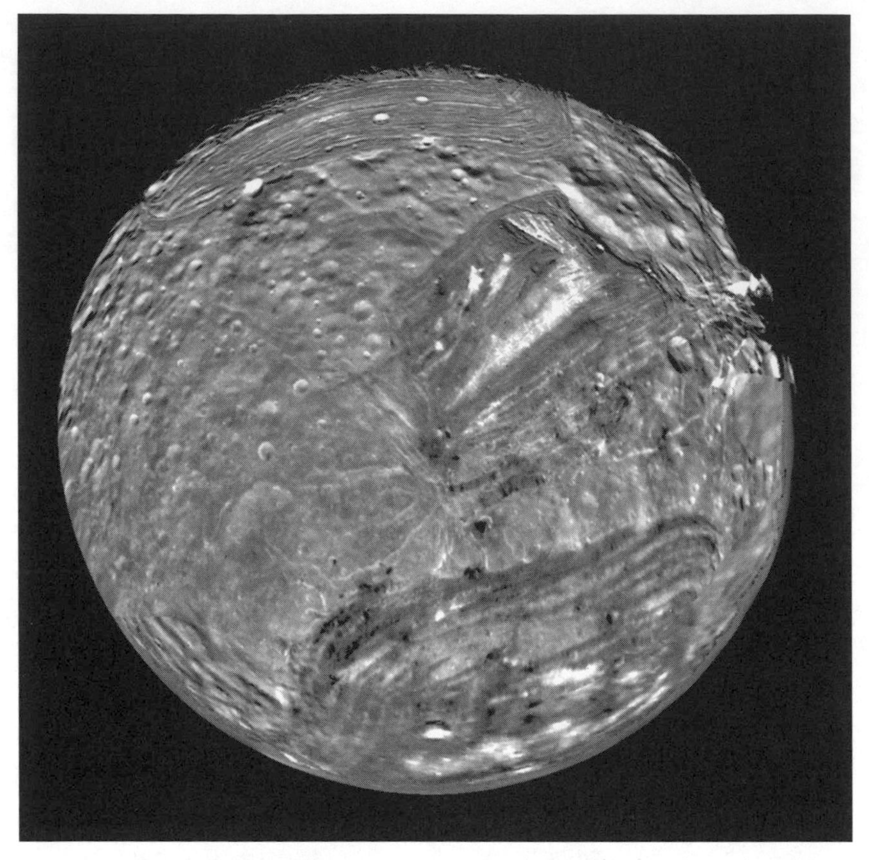

Figure 8.3. *Voyager 2* Image of Uranian Moon Miranda, 1986. Uranus has at least fifteen satellites, ten of them discovered by *Voyager 2.* Miranda, shown in this nine-image mosaic, is the most surprising because of its geologic diversity. On the surface of this small, icy moon, only 185 miles in diameter, giant canyons as much as twelve miles deep slice across heavily cratered plains that abut large patches of grooved and ridged terrain. Miranda appears to have been frozen at an early stage of its development, and reveals a perplexing surface unlike any other known to exist in the solar system. NASA/JPL.

mid-1990s. Therefore, it is not considered reasonable for NASA to undertake making up the entire deficit by the necessary increase in DSN aperture. It would require more than doubling the present aperture, an approximate $200 million enterprise." So where could the DSN turn to resolve the conflicts imposed upon it by the *Voyager 2* Neptune encounter? In typical Stevens fashion, he already had an innovative plan, an essential part of which circumvented the need for new antennas by enhancing those that already existed in the network. He had set forth his findings in a lengthy, and highly technical, document titled "Implementation of Large Antennas for Deep Space Mission Support" (Stevens 1984).

The DSN proposed to support the *Voyager 2* Neptune encounter in 1989 by enhancing the resources used for the Uranus encounter in three unique ways. First, the existing 64-meter antennas would be enlarged to 70-meter diameter. Second, several large non-NASA antennas—Parkes in Australia and the Very Large Array in Socorro, New Mexico—would be arrayed with the DSN 70-meter antennas. Third, all of the DSN 34-meter antennas, including those of newer, high-efficiency design, would also be arrayed with the 70-meter antennas. Communications link design calculations showed that, in combination, these resources could make up 3.2 dB of the 3.5-dB deficit created by the increased distance to Neptune. Taking account of other factors related to the Australian (Southern Hemisphere) view of the actual encounter, that was deemed close enough.

It was, in fact, a brilliant scheme that took advantage of the Neptune encounter requirements to justify the enlargement of the 64-meter antennas, a goal that since the very beginning had excited the interest of the forward-thinking DSN engineers. In addition to its application to the *Voyager 2* Neptune encounter, the proposed 70-meter antenna would extend the useful life of other spacecraft, notably *Pioneer 10,* which was then exiting the solar system and returning data from that scientifically interesting region of the heliosphere. By reducing the need for multiantenna arrays, it released the 34-meter antennas to provide additional support to other space missions, where availability of antennas rather than threshold signal levels was the constraining factor. Radio astronomy and planetary radar experiments would also benefit from the additional sensitivity for detection of weak signals. Since the increase in antenna gain that resulted from the increased aperture applied as much to the uplink as to the downlink, the effective radiated power of the DSN uplink transmissions was also increased, a feature that could be applied to economize in the cost of primary power for ground facilities, or used effectively to revive ailing spacecraft where attitude control problems had mispointed the spacecraft's own antennas. This latter feature would prove to be of inestimable value many years later on *Galileo*'s mission to Jupiter, as we shall see.

Nevertheless, as Stevens knew only too well, it was one thing to develop a brilliant plan at JPL and quite another to persuade NASA headquarters to approve and fund it. For that purpose the DSN needed a powerful advocate, someone to fill the role that Rechtin had played twenty years earlier for the 64-meter antenna. The role fell to Peter Lyman.

The Proposal

It began early in 1982 with a long and technically challenging memorandum from Robertson Stevens, then Lyman's chief engineer, that addressed the deteriorating condition of the network's three 64-meter antennas, the current demands upon them, and their future prospects. Stevens pointed out that all three antennas had by then been in continuous operation for a very long time, the Goldstone antenna for nearly sixteen years, the antennas at Madrid and Canberra for nearly nine. The original design, developed twenty years earlier, had specified a working life of ten years for the mechanical parts that were subject to wear. Now replacement components were becoming difficult to find, or costly to fabricate. Moreover, during its years of service, engineers had identified many deficiencies in the original antenna system design and implementation (Wood 1993).

With many existing spacecraft pursuing extended missions at ever-increasing distances, and with more ambitious missions being proposed for the distant future, antennas with greater effective collecting area (aperture) and higher operating frequency (such as X-band) were required.

Stevens observed that the 64-meter antennas would be a vital NASA and national technical facility resource for many years thence, and proposed a way to remedy the increasingly serious situation then confronting the network. There were two major parts to his recommendation, the first concerned with the deteriorating components and structure of the antenna, the second with an increase in performance.

As we have seen in previous chapters, the problem of oil film height that had appeared at the Goldstone antenna early in its operational life had continued to plague Dale Wells and his mechanics. Constant shimming and regrouting consumed manpower and antenna time that would otherwise have been available for spacecraft tracking support. Since this had not occurred at the overseas antennas, it seemed that the Goldstone pedestal had a unique problem. What could possibly be wrong with the high-test concrete so carefully placed by Casperson, Phillips, and McClure that cold winter morning at Goldstone so long ago? Stevens proposed to find out.

There were other problems, too, which were not confined to the Goldstone antenna. These included the azimuth radial bearings, the master equatorials, and the huge gearboxes that drove the pinion gearwheels around the azimuth and elevation bull gears.

Stevens proposed to set up a permanent Technical Facilities Board, comprised of JPL's best engineering talent and empowered to bring in specialist consultants from anywhere in the United States or the world, to provide expert advice on which to develop a rehabilitation plan. It would be up to Lyman and Johnson to persuade NASA headquarters to support the plan with the appropriate level of funding.

Next Stevens turned to enhancing the performance of the existing 64-meter antennas, basing his approach on an earlier study by J. R. Hall. The study (1975) showed convincingly that it was practicable to increase the diameter of the existing antennas to 70 meters, and to extend the operating frequency of such antennas to X-band and possibly even to the Ka-band region.[5] It predicted substantial improvement in the radio frequency performance of the antenna as a result of such modifications.

There was much more to it than merely adding additional panels to the existing reflector surface. The desired enhancement could be achieved only by completely replacing the existing reflector surface with a larger surface of considerably greater refinement. Recently developed holographic techniques would be employed to set the new antenna panels to the desired surface contour more accurately than had been possible in the past. A specially shaped subreflector would be necessary to illuminate the main reflector in an optimal way, and improved microwave and thermal-noise-reduction techniques would be required to achieve the desired radio-frequency figure of merit.[6]

For upgrading the existing 64-meter antennas, Stevens presented three options, which he called basic, intermediate, and ultimate. The basic option called for further strengthening of the tipping structure to compensate for slight distortions due to gravity effects when the antenna moved in the elevation direction, an important consideration for operating at X-band or Ka-band but of less consequence at S-band. The cost in 1982 dollars was estimated at $0.33 million per antenna for an increase in X-band performance of 0.3 dB (a factor of 1.07). The intermediate option added a shaped subreflector to improve the focusing of the microwave energy from the main reflector surface into the ultrasensitive receiver horns that were mounted on the apex of the tricone.[7] The cost increased substantially to $3.3 million per antenna, but so did the performance, by 1.4 dB (a factor of 1.38) at both X-band and S-band.

The "ultimate" option pushed the art of X-band antenna design to a new level. It called for even more efficient illumination of the main reflector by requiring much greater precision in the design, construction, and alignment of the shaped surfaces of both the subreflector and the main reflector. With compensation for gravity-induced distortion, Stevens predicted a performance increase of 1.9 dB (a factor of 1.55) at X-band. The S-band performance would remain as before, but the cost now jumped to $6 million per antenna.

As key elements of the microwave system, precision-shaped subreflectors were costly elements indeed. Stevens drove this home by pointing out that the three X-band options under consideration would effectively cost NASA $1.1 million, $2.4 million, and $3.2 million per decibel of improved performance (factor of 1.25) respectively. "By way of comparison," Stevens wrote, "the addition of one new 34-m antenna to an existing 64-m antenna to form an 'array' would cost $6.0 million, for an effective cost of $4.6 million per decibel at X-band." Insofar as the "best X-band bang for the buck" was concerned, there was no contest, but could NASA be persuaded to support such a proposal? Stevens thought so. His March 1982 report to Lyman concluded, "I recommend that we embark on a program to design and implement the full technical performance potential of the 64-m antennas. That is the 'Ultimate' configuration, with the predicted performance and costs [as presented]" (Wood 1993).

Under Lyman's forceful direction, the Tracking and Data Acquisition Office at JPL lost no time in conveying Stevens's recommendations to NASA headquarters. Six months later the 64-Meter Antenna Rehabilitation and Upgrade Project was funded, staffed, and under way. The project manager would be Don McClure.

The Plan

Don McClure ended his close association with the Goldstone 64-meter antenna right after it went into operational service in 1966. For the next decade he gradually moved up through various JPL technical organizations associated with the Deep Space Network. By mid-1982, when the TDA Office decided to proceed with the 64-meter antenna upgrade project, McClure's demonstrated success in the execution of large DSN-related implementation tasks and his unique on-the-job experience in the construction of the Goldstone 64-meter antenna made him an obvious choice for manager of the new project. His low-key, nonconfrontational style of project management had proved of great value on large construction projects such as this, which crossed organizational boundaries at both JPL and NASA headquarters and often involved arrangements with NASA's international partners.

McClure now found himself in a situation very reminiscent of Merrick's twenty years earlier. This time, though, McClure was the project manager and, like Merrick, his first action was to enlist the services of an experienced, dependable "first lieutenant." Many of the former members of Merrick's Hard Core Team were still at JPL, although over time they had been dispersed throughout the organization. Among them was Fred McLaughlin, and it was he that McClure tapped for the role of project engineer. Together they developed a plan of action.

The work would be carried out in two phases. First they would deal with the big concrete repair jobs on the hydrostatic and radial bearings at Goldstone. That would take DSS-14 out of service for six to nine months, during which time the spacecraft tracking functions it normally handled would be carried by the smaller, 34-meter antennas. It was, of course, essential that all of this work be completed on schedule to permit the rehabilitated 64-meter antenna to cover *Voyager 2*'s Uranus encounter in January 1986.

In the second phase of the work, the three 64-meter antennas would be upgraded to meet the new 70-meter performance specifications. However, the schedule on which the work could be accomplished was driven by the powerful voices of the space flight missions, particularly of *Voyager 2* and *Galileo*. *Voyager*'s requirements for the new 70-meter antenna were fixed by its 1989 arrival date at Neptune, and could not be changed. *Galileo* turned out to be the constraining factor. With a launch in June 1986, the mission timeline allowed only an eighteen-month window of opportunity in which to complete the upgrade work.

When the project began, McClure estimated that it would take twelve months to modify a single 64-meter antenna to 70-meter diameter, or three years to do all three consecutively (McClure 1999b).

It was a clear impasse. Either two antennas would have to be out of service simultaneously at some point, or modification of one of the three 64-meter antennas would have to be delayed. Neither the Galileo nor the Voyager people would agree to have two antennas out of service at the same time, and Voyager preferred that the Madrid antenna be delayed, while Galileo preferred that Goldstone be delayed. McClure was left to sort it out. He decided to get the work started and defer resolution of the two-antenna conflict to later. It turned out to be a wise decision, for reasons that were as unfortunate as they were unforeseen.

The matter was decided for McClure in January 1986 when the space shuttle *Challenger* was lost and the launch of *Galileo* on its mission to Jupiter was delayed indefinitely. The eighteen-month constraint was removed, and only the *Voyager 2* encounter with Neptune in August 1989 remained as a controlling milestone. The implementation schedule was reworked to take advantage of the relaxed conditions, although McClure set June 1, 1988, as the completion date for the third antenna. The essential elements of the final plan are shown in figure 8.4 (McLaughlin 1999).

From the outset McClure recognized that meeting the preset antenna downtime schedule was critical to the success of his project. To cut the actual downtime to an absolute minimum, he developed a detailed erection plan based on the assumption that most of the assembly work could be carried out on the

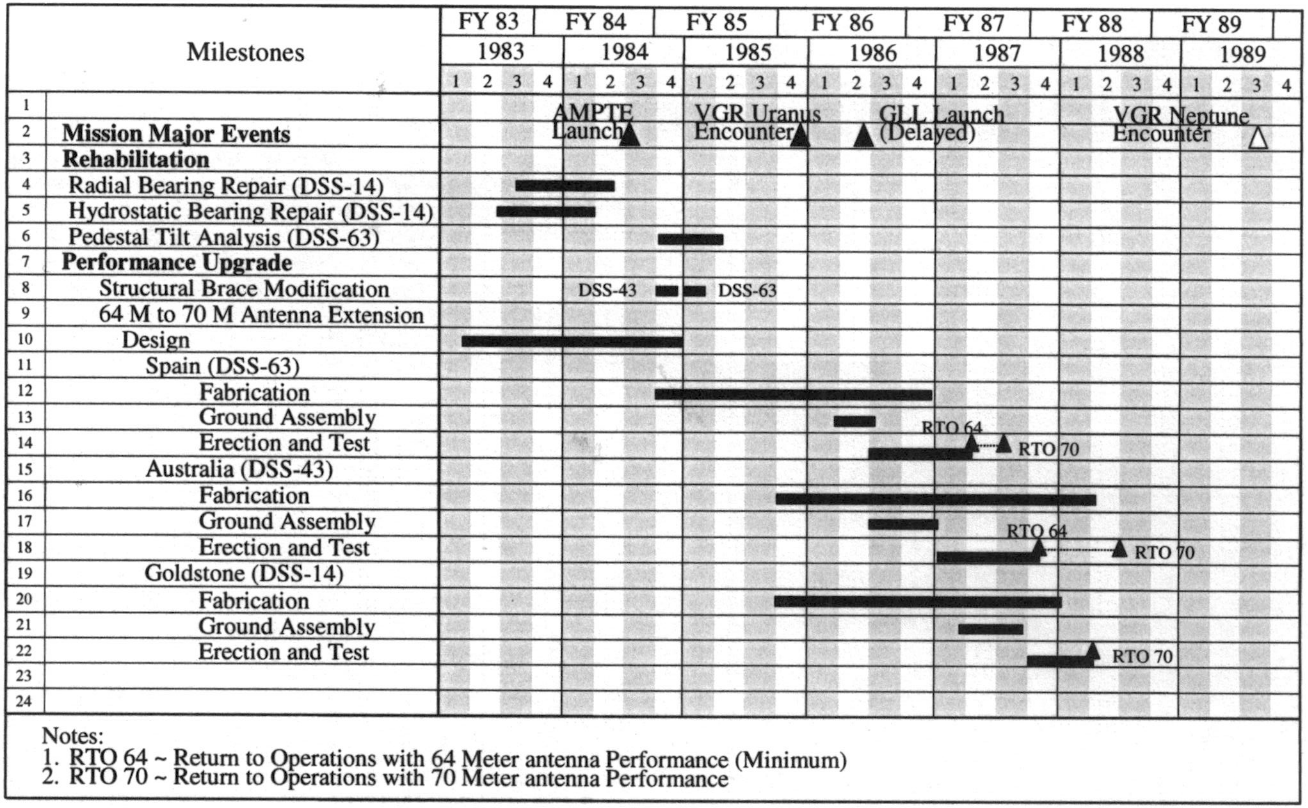

	Milestones	FY 83 (1983)	FY 84 (1984)	FY 85 (1985)	FY 86 (1986)	FY 87 (1987)	FY 88 (1988)	FY 89 (1989)
		1 2 3 4	1 2 3 4	1 2 3 4	1 2 3 4	1 2 3 4	1 2 3 4	1 2 3 4
1			AMPTE Launch	VGR Uranus Encounter	GLL Launch (Delayed)		VGR Neptune Encounter	
2	**Mission Major Events**							
3	**Rehabilitation**							
4	Radial Bearing Repair (DSS-14)							
5	Hydrostatic Bearing Repair (DSS-14)							
6	Pedestal Tilt Analysis (DSS-63)							
7	**Performance Upgrade**							
8	Structural Brace Modification		DSS-43	DSS-63				
9	64 M to 70 M Antenna Extension							
10	Design							
11	Spain (DSS-63)							
12	Fabrication							
13	Ground Assembly							
14	Erection and Test					RTO 64 RTO 70		
15	Australia (DSS-43)							
16	Fabrication							
17	Ground Assembly							
18	Erection and Test					RTO 64	RTO 70	
19	Goldstone (DSS-14)							
20	Fabrication							
21	Ground Assembly							
22	Erection and Test						RTO 70	
23								
24								

Notes:
1. RTO 64 ~ Return to Operations with 64 Meter antenna Performance (Minimum)
2. RTO 70 ~ Return to Operations with 70 Meter antenna Performance

Figure 8.4. Planning Schedule for 64-Meter Antenna Rehabilitation and Performance Upgrade Project, 1983. The Ground Assembly bar shows where prefabricated components were assembled into modules on the ground for later installation on the antenna. The Erection and Test bar shows the antenna "downtime," where the 64-meter elements were removed and replaced by 70-meter components, and the period where the upgraded antenna was brought up to the specified 70-meter performance level. Illustration by Fred D. McLaughlin, NASA/JPL.

ground, before work began on the antenna itself. The inner and outer rib trusses, for instance, could be assembled into modules well ahead of time. Each of the sixteen assembled modules would represent a 22.5-degree segment of the complete reflector surface (McClure 1999b).

Work on the antenna itself would begin with the removal of individual surface panels from the center section, out to a distance of 34 meters. The remaining panels, with backup trusses attached, would then be removed as complete modules. This made for heavy lifts, but it would save antenna downtime. Each of the twelve modules was estimated to weigh about 10 tons, and they were to be lifted off the center section by rotating the antenna under the boom of the heavy-duty crane in an opposing sequence to symmetrically balance the dead-weight loads on the remaining structure.

In the next step, the center section would to be strengthened to carry the extra weight of the new antenna backup structure, quadripod, and subreflector, and the increased counterweights.

Completion of this step would initiate the actual reassembly sequence, starting with the two counterweights, one for each elevation wheel. Each counterweight weighed 140 tons[8] and would be built up incrementally from heavy rectangular steel plates in such a way that the whole structure was kept in balance. Then the sixteen preassembled inner rib trusses were to be attached and aligned. Concurrently the new quadripod was to be assembled on a special foundation on the ground and held ready to be lifted into place as soon as the old quadripod and subreflector were removed.

Installation of the sixteen outer rib modules would follow basically the same procedure. When all modules had been aligned, the intermediate ribs, hoops, and diagonal braces would be installed to complete the reflector backup structure.

After they were assembled and tested at ground level, the subreflector and its positioner would be lifted by the quadripod hoist to the apex of the quadripod and fixed permanently in position.

The surface panels would be the final item in the erection plan. Once the entire supporting structure was completed, these panels were to be installed one by one, starting from the innermost row. They had to be handled with great care to avoid scratching the surface or otherwise damaging them, and they were to be set very accurately, with gaps the thickness of a credit card between them.

After all 1,272 panels were installed, the overall surface would be set to achieve the desired contour shape at a 45-degree elevation angle with an accuracy of 0.025 inches (0.6 mm), again about the thickness of a plastic credit card.[9] Because the manual adjustments to individual panels had to be carried out with the antenna pointing vertically, a structural analysis program was used to calculate the offsets to be applied to each adjustment to compensate for the gravity sag when the antenna was set at the desired rigging angle.[10] Alignment of the subreflector and calibration of the intermediate reference structure were the final items in the erection plan.

At this point the erection portion of the overall project would be essentially finished. The 70-meter antenna would be complete in all respects except for the RF measurements to confirm that it met the specifications for gain and noise temperature. Finally, a series of tracking passes with an in-flight spacecraft would be required to demonstrate that the 70-meter antenna was ready to return to full support of mission operations.

This, then, was McClure's plan, but a lot of engineering was needed before it could be put into effect. That task fell to Floyd Stoller.

The Design

Manager Floyd Stoller passed the word down to the lead engineers in his Ground Antennas and Facilities Engineering Section with a certain degree of personal satisfaction. He was now responsible for the engineering design of the enhancements to the Big Dish, the largest enterprise of its kind that the DSN had undertaken since Merrick directed the original engineering design work twenty years earlier. Six months previously, the formation of the 64-Meter Antenna Rehabilitation and Upgrade Project with Don McClure as project manager had been announced. A plan and schedule had been developed, and by March 1983 Stoller was ready to assign the design work for which he was responsible to his engineers, each a specialist in his own field. Many of the engineers from the original Hard Core Team were still at JPL, and the opportunity to extend the performance limits of their original designs constituted a challenge that none would pass up.

Stoller began by addressing the basic issue of structural integrity. While it was obvious that an increase in the diameter of the antenna would be attended by an increase in weight, it was not known how much the weight would increase, or how well the elevation bearings, the hydrostatic bearing and radial bearing, and the drive system would be able to handle it. Stoller assigned Houston McGinness to find out. It took him about a year, but when he was done his conclusions were very reassuring.[11]

The study reviewed the original reasoning that determined the size of the two giant elevation bearings for the 64-meter antenna and the structure that supported them, and found it still valid for the 70-meter design. They could carry the additional weight without any changes. Furthermore, the study showed that the all-important oil film height on the hydrostatic bearing would not be significantly affected by the additional weight, nor would the stability of the structure be driven beyond its design limits by the increased weight and wind loading. Analysis of the additional stresses on the radial bearing and its component parts showed that they too were well within acceptable working limits. Finally, McGinness concluded that, despite the increased torque needed to drive the larger-diameter antenna against the wind, the torques available in the existing 64-meter drive systems were adequate for the 70-meter antenna. There was, in fact, no need to make any changes to these four basic mechanical systems in order to increase the diameter to 70 meters, a conclusion that was a fine tribute to the foresight of the original designers.

In yet another critical area of the 70-meter design, the tipping structure, Stoller turned to a specialist in the complex mathematical field of finite element analysis, Smoot Katow. Like McGinness, Katow had been heavily involved in analyzing the behavior of the tipping structure of the 64-meter antenna. Now he was asked to use the same analytical method to determine the behavior of the deeper trusses and hoops that would replace the original assemblies to extend the surface of the main reflector out to a diameter of 70 meters. It was vital to verify that the labyrinth of steel trusses, hoops, girders, and other structural members that were needed to support the mammoth new antenna surface would not deflect under the effect of the wind, or sag under the antenna's own weight as the pointing position tipped from horizontal to vertical. More precisely, it was important to know how much they would deflect, for in either case such distortion would adversely affect the precision-shaped surface of the main reflector and degrade its pointing and/or microwave performance. The mathematical model that Katow developed was too large to fit on the mainframe UNIVAC computers that were available at that time, and a compromise solution was finally used to run the computations. Nevertheless, Katow was satisfied that the increased cross-sectional areas of the new trusses and the addition of reinforcing plates to critical areas of the existing inner rib trusses would create a tipping structure of sufficient stiffness to fall within the tipping structure's share of the overall error budget.

To complete the mechanical and structural design work for the 70-meter upgrade, Stoller looked to two of his senior engineers, Bob Hughes and John Cuchissi. Neither had been involved with the original Hard Core Team, but they had been heavily involved with all of the DSN antennas for many years since then and were eminently capable of designing the new subreflector positioner and the new quadripod respectively.

The subreflector positioner acted as a kind of focusing mechanism that automatically adjusted the position of the subreflector to compensate for gravity effects at the apex of the quadripod as the antenna moved through its full range of tipping motion. A new and considerably stronger positioner was required, because the new subreflector was heavier and considerably larger than the existing subreflectors on the 64-meter antennas and was therefore subject to much higher wind loads. Hughes's design succeeded in providing the necessary extensions in operating range and torque, while at the same time adding new features, such as Z-axis tilt capability, for future use.

John Cuchissi faced a different design problem—how to make the four legs of the quadripod longer and thinner, yet stronger, than those on the 64-meter antenna. While the primary purpose of the quadripod was to support the subreflector and the positioner, it also served as a crane for lifting the subreflector and feed cones and the heavy pieces of electrical and microwave equipment that were mounted in them. This meant that the quadripod had to be capable of lifting the 13,000-pound subreflector, it had to be stiff enough to minimize the deflection due to gravity as the antenna moved in elevation, and it had to present the smallest possible shadow area to the microwave energy reflected off the antenna surface. It had, in other words, to be strong, rigid, and thin.

In cross-section the legs were roughly half the size of those on the 64-meter antenna. And they were longer, 93 feet versus 89 feet. It seemed a very flimsy structure indeed to support the huge weight of the subreflector and positioner at its apex without collapsing, let alone merely deflecting slightly as the antenna moved from horizontal to vertical. There was no precedent for a design such as this, and it seemed prudent to verify the design performance before proceeding. To do this, a full-scale model was built and subjected to exhaustive tests to verify the design parameters and the margins of safety over a catastrophic failure due to buckling of any of its components. Needless to say, Cuchissi's design satisfied the test criteria and no further changes were necessary.

In addition to the structural, mechanical, and facilities engineering, there was another crucial element of the project—the design of the main reflector, subreflector, and microwave optics systems. This lay in the field of microwave engineering, an area covered by Dan Bathker, the engineer who had been closely associated with the original microwave design for the 64-meter antenna. Working out of the Antenna Microwave Section with a small group of expert microwave engineers, Bathker assumed responsibility for the design of the 70-meter antenna microwave optics system and answered to the project manager for its design integrity, fabrication, installation, and overall performance validation.

"Radio frequency optics" was the term Bathker used to describe his special area of expertise in the general field of microwave antenna engineering, and it was there that the answers to the last of the design issues for the new 70-meter antenna were to be found. The techniques used to design and build the microwave feeds, or illuminators, and the specially shaped surfaces of the primary reflectors and subreflectors included a modified form of the ray-tracing techniques used by optical engineers to design telescopes and lenses. Microwave engineers use the term "RF optics" to distinguish the former from the latter. The design of the Cassegrainian RF optics system for the 70-meter antenna was determined by the size and shape of the antenna, the size and shape of the subreflector, and the feed mounting position on top of the existing tricone.[12] In actuality it was an interactive procedure between the design of the feed horn and the ultimate design for shape and size of the antenna and subreflector. In the trade-off studies that ensued, factors of cost, availability, performance, and potential for future improvement were all considered in reaching a decision on this critical element of the RF optics. The ultimate choice was the "22-dBi feed horn,"[13] a device that had been developed some years earlier in Bathker's lab and optimized for use on X-band only. Although it had been tested on the 64-meter antenna, its full X-band performance potential could not be realized on that antenna without refining the antenna surface. However, for the 70-meter surface it was a perfect match. And so it was decided—the 70-meter antenna would be designed to match the RF optics of the 22-dBi feed horn.

The surface panels, too, were a critical factor in this decision. The radio frequency performance of a microwave antenna is a function of the "smoothness" of the reflecting surfaces. Microwave engineers obtained a qualitative measure of the smoothness by measuring a number of points on the surface to determine the errors with respect to the desired surface shape or contour. The errors were measured in a direction normal (at right angles) to the surface. The root mean square (RMS) of these measurements was then used to specify the surface smoothness, or accuracy, of the finished surface. For individual panels, a measurement was made for every ten square inches of panel area; for the surface of the main reflector, measurements were made at each corner of the individual panels.[14]

In the early 1960s when the 64-meter antenna was designed, the main reflector surface used panels of the highest precision then available. Bob Hall had built them down at Rohr by riveting a stretch-formed sheet of aluminum to a contoured metal framework. The specified accuracy for individual panels was 1.52 mm (0.06 inches), and Rohr had done better than that. Nevertheless, manufacturing techniques had improved over the intervening twenty years, and by 1984 it was possible to build panels with an accuracy more than ten times better. Now Bathker wanted to specify panels with a surface accuracy of less than 0.1 mm (0.004 inches). This improvement in surface accuracy, he claimed, would increase the gain of the 70-meter antenna by 0.5 dB (a factor of 1.12).

But there was a catch, as McClure was quick to point out, and that was the cost.

The surface area of the new 70-meter antenna was just over one acre (0.4 hectares) or 45,500 square feet (4,227 square meters), equivalent to four standard homesites. It required 1,272 panels of seventeen different sizes, and for Don McClure, who controlled all the expenditures on the project, the problem was to find a panel design that would meet Bathker's accuracy goal while at the same time limiting the cost of fabrication to $75 per square foot of panel surface.

After investigating several methods of fabricating panels, McClure and McLaughlin finally chose a bonded metal technique as the best approach for this particular application. Adhesive-bonded aluminum was widely used in the aircraft industry, and reliable data on strength, life expectancy, and environmental effects were readily available. Furthermore, tests at Goldstone had shown that, compared to riveting, the bonding technique produced panels with better microwave performance at X-band and significantly better performance at Ka-band. And finally, a contractor was found that could meet the target cost of $75 per square foot.

The enhanced performance that accrued from the use of shaped subreflectors to illuminate Cassegrain-type microwave antennas was well known to JPL antenna engineers, and designs of that type had been used and tested on several smaller DSN antennas in the past. Determined to wring every fraction of a decibel of performance out of the new 70-meter surface, Bathker elected to specify a precision-shaped subreflector to illuminate it, presenting McClure and McLaughlin with yet another challenge in its fabrication—one that almost got the better of them.

The increased surface area of the new antenna, and the fact that it too was to be "shaped" rather than parabolic like the 64-meter antenna, established the need for a totally new subreflector with a surface accuracy to match that of the main reflector. As if that was not enough, the offset microwave feed arrangement on the new antenna required the surface of the subreflector to be asymmetrical rather than a straightforward surface of revolution. The specification was, consequently, very demanding indeed. A thorough analysis of the factors involved in trading off diameter against performance, weight, and cost resulted in the decision to make the diameter 25.6 feet (7.8 meters) and the weight 13,600 pounds (6,168 kilos).[15]

An evaluation of the options for fabrication favored a cast-aluminum blank that would be precision-machined to achieve the desired complex contour with a surface accuracy of 0.01 inches (0.25 mm). This, however, was not the end of the matter. The decision to cast the subreflector raised new questions. As McClure pointed out, "Casting an object [this large and this heavy] represented a difficult casting challenge. Should it be cast in one piece? If not one piece, then in how many pieces, and in what shape? How would the pieces be joined? Would they be welded or bolted?" (McClure 1999b).

After wrestling with these and similar questions for the best part of a year, McClure and McLaughlin eventually decided "to build the subreflector as a two-piece unit in which each half consisted of three separately cast pieces that were welded together. The two halves were designed to bolt together to permit the unit to be machined as a single piece. In this way, the subreflector could be shipped in two pieces and assembled at the site prior to installation" (McClure and McLaughlin 2001).

Bathker and his colleagues supplied the final two defining characteristics of the subreflector surface—finish and porosity. Based on microwave measurements they had conducted at JPL, Bathker called for a finish of 125 millionths of an inch. That was a very smooth finish indeed, and would require a very special digitally controlled machine to shape the raw casting. The porosity related to the inevitable blowholes that are a natural feature of large castings. Holes larger than a quarter inch in diameter were to be filled with weld and ground smooth. Smaller holes were not detrimental to the performance of the subreflector and were to be filled with epoxy and sanded smooth.

We turn now to look at how McClure's schedule and erection plans, Stoller's engineering designs, and Bathker's exquisite microwave optics were implemented at the three far-flung sites around the world and how, even in the DSN, the best laid plans may often go astray.

9 Implementation

Aggravation in the Aggregate

In 1999, more than thirty-three years after he poured it in and almost sixteen years after he gouged it out, Don McClure still retained a lasting respect for the concrete in the pedestal of the Goldstone 64-meter antenna. "For concrete that was allegedly no good, it was incredibly difficult to get out," he said. Of course, he did neither job personally, but in both cases he had been JPL's man on the job, responsible for seeing that the work was properly carried out.

When JPL entrusted him with the rehabilitation of the Goldstone 64-meter antenna pedestal in 1982, McClure was fully confident of handling the unusual task successfully. On his team at JPL he had engineers like Houston McGinness, Horace Phillips, and Verl Lobb, experts from way back who had been on Merrick's original Hard Core Team that built the antenna. At Goldstone he had talented, practical engineer Dale Wells to supervise the actual work. McClure appreciated Wells's unique gifts. "Dale had a remarkable talent to talk intelligently with engineers, technicians, and upper management. He could stand up in front of JPL or NASA top brass and explain the complexity of the hydrostatic bearing and pedestal in words that everybody could understand. Not everyone can do that. JPL was very fortunate to have a man like him," he said. Furthermore, as the DSN engineer responsible for maintenance of the electromechanical systems, Wells had been continuously associated with the Goldstone 64-meter antenna since 1966, when Merrick first brought it into service. If anyone knew its strengths, weaknesses, and idiosyncrasies, it was Dale Wells (McClure 1999a).

As explained earlier, the rotating section of the antenna was supported by a cylindrical reinforced-concrete pedestal. It was 80 feet in diameter and 30 feet high. Over the years, the weight of the antenna structure had increased from 5 million to about 8 million pounds with such additions as the tricone, subreflector, high-power transmitter, and the extra counterweights required to balance them. The vertical load of the antenna was supported by three hydrostatic bearing pads, spaced equidistantly around the perimeter of the pedestal. Between each pad and a five-inch-thick circular steel runner, a thin film of oil provided the hydrostatic bearing surface upon which the pads rode around the runner. The runner rested directly on steel sole plates that were attached to the top surface of the pedestal and precisely leveled by means of specially formulated grout packed between its bottom surface and the rough upper surface of the concrete pedestal. The essential features of the pedestal and the hydrostatic bearing are illustrated in figure 9.1.

The original grout that Merrick's engineers specified for the packing between the sole plate and the pedestal was a special nonshrinking product designed to permanently maintain the extreme flatness required by the runner. Unfortunately, as we have seen, with the passage of time the grout material reacted with the steel sole plates under the runner to form rust, which caused the runner to lift unevenly and thereby destroyed its precisely flat surface. By 1969, just three years after Merrick put the Goldstone antenna into service, the grout had deteriorated to the point where it had to be replaced. Horace Phillips was given the job of replacing all of the grout with an improved Portland cement product. That, however, was not the end of it. Engineers were soon back at the routine task of inserting shims under the sole plates to keep them level as the grout apparently sank beneath the tremendous weight of the bearing pads. By 1976 the grout problem had become so serious that JPL initiated an in-depth study of polymer concrete, cement grout, chemical resistance to oil, and other aging characteristics to try to determine the cause, and cure, of the problem. Some of the world's foremost authorities on concrete technology were called in to study, test, and advise JPL on how to correct this apparently intractable problem, and to ascertain why it had affected only the Goldstone antenna.

The investigations were costly, time-consuming, and extremely detailed, and the resulting reports (see McClure 1985) were complex. The bottom line, however, was clear enough. It was not the grout that was deteriorating but the pedestal concrete. Why? Because the aggregate for the concrete, which had been so

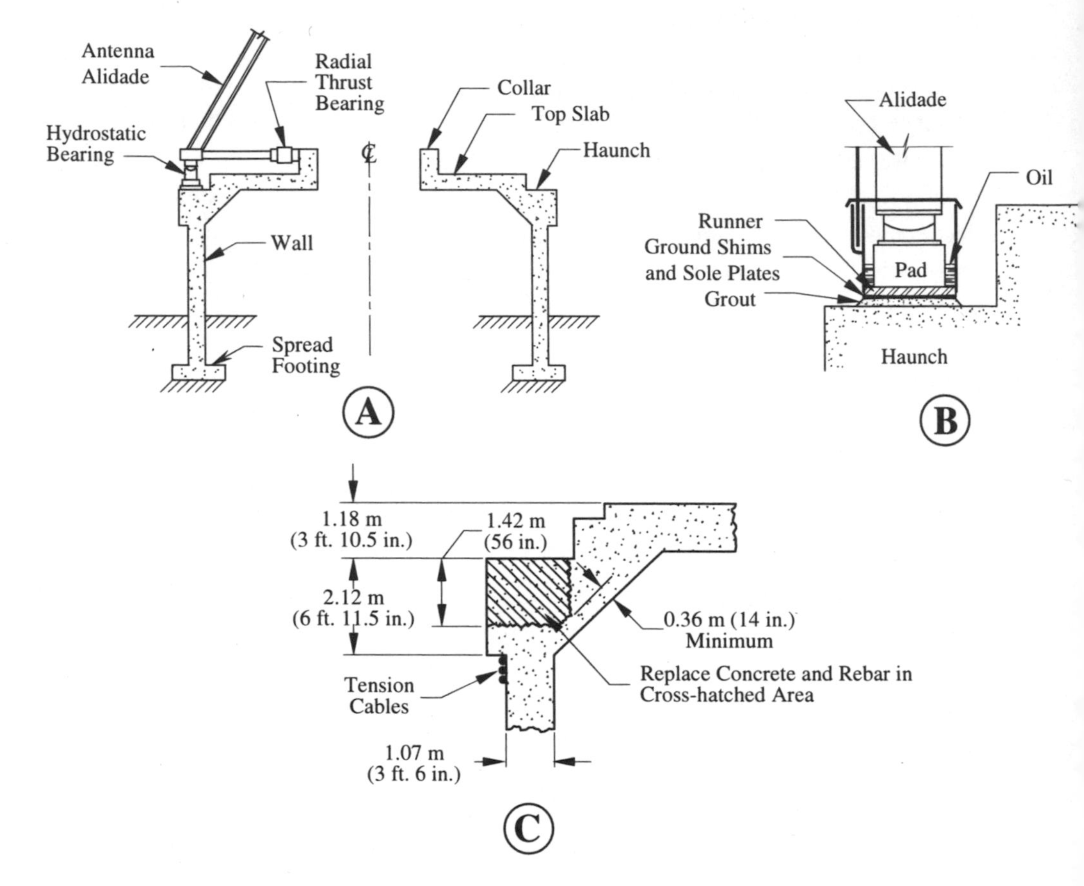

Figure 9.1. Diagram of Pedestal, Hydrostatic Bearing, and Haunch Concrete. In this simplified cross-sectional representation of the pedestal, the first diagram (A) shows the wall that supports the vertical weight of the antenna by means of the hydrostatic bearing, and the collar that supports horizontal loads, such as wind effects, by means of the radial thrust bearing. The middle diagram (B) shows the arrangement of grout, sole plate, and runner on top of the haunch section of the pedestal wall. To avoid "grounding out" the pads on the runner surface, it is essential that the correct oil film thickness be maintained at all times, a requirement that depends directly on the permanent stability of the grout under the sole plates. The lower diagram (C) shows the haunch concrete under the hydrostatic bearing where the haunch concrete had to be removed and replaced, and the circumferential cables that ensured the integrity of the structure while the work was carried out. Illustration by Fred D. McLaughlin, NASA/JPL.

carefully selected by Merrick's engineers for its strength and stability, had reacted with the alkali in the cement to form silica gel.[1] The natural tendency of silica gel to absorb moisture caused it to expand, and in so doing to create micro cracks in the surrounding concrete. These millions of tiny cracks in the concrete near the hydrostatic bearing adversely affected its elasticity,[2] or ability to conform to the hydrostatic bearing and runner deflections under load. The evidence was compelling—the concrete would have to be replaced.

Why was this unique to Goldstone? It was only at Goldstone that an unfortunate choice in the source of supply had led to the use of reactive aggregate for the concrete mix.[3]

All this was behind him when McClure accepted the manager's job for the 64-Meter Antenna Rehabilitation and Upgrade Project in September 1982. Now, in mid-1983, the rehabilitation portion of the project lay immediately before him, and the window of opportunity, the nine-month period when the 64-meter antenna could be taken out of operation, was fast approaching. For the past several months, Dale Wells had been working with a civil engineering consulting firm to figure out how to support the antenna structure while they gouged out the concrete under the hydrostatic bearings on which it rotated. Whatever scheme they used would have to support 6 million pounds of dead weight for the best part of a year in all weather and wind conditions, and be able to survive a modest earthquake without moving more than the smallest fraction of an inch. There could be no question of risk to the workers or of damage to the antenna itself.

In the final plan, three massive steel columns, each one solidly bedded in a reinforced concrete foundation, would be erected at the corners of the alidade to carry the weight of the antenna after the pads, runner, and sole plates had been removed. McClure also planned to refurbish the radial thrust bearing while the antenna was down for the hydrostatic bearing work. To hold the antenna from moving sideways while Wells worked on the radial bearing that normally held it in position, the corner weldments of the alidade were securely attached to the center collar by big I-beams running across the base triangle.

The plan looked great on paper, but would it, literally, hold up? When someone questioned how much the "stilts" would settle under all that weight, Wells arranged a practical demonstration. He rented the biggest and heaviest mobile crane he could find, took it out to the antenna site, and put a one-foot-square steel plate under each of the four jacks. Then he jacked it up so the pads supported the full weight of the crane and set up a theodolite nearby to measure the settling over a four-week period. On the basis of these empirical data, he estimated that the foundations of the "stilts" would settle about three-eights of an inch. He thought he could live with that.

Meanwhile, hydraulic jacks had been placed under each corner of the alidade to give a total lifting capability of 2.8 million pounds at each corner. The lift began on June 13, one-tenth of an inch at a time. As a safety measure, heavy steel "stools with shims" followed the extension of each jack in case it failed. The antenna structure had to be raised just enough to allow removal of the pad assembly. It took most of the day to complete the desired one-inch lift, at which point the weight was slowly transferred to the "stilts" over a three-day period to allow the foundations to gradually adjust to the load. After initial settling of a little less than half an inch, no further movement was detected, and the runner segments were quickly removed to allow access to the actual concrete. The diagram in figure 9.1 highlights the portion of the haunch where the concrete was to be removed and then replaced.

Although the concrete would be removed only in three small, equally spaced sections at a time, and the voids would be filled with new, cured concrete before work moved to adjacent sections, Wells was concerned about the structural integrity of the entire pedestal, particularly when the jackhammers went to work on the old, tough, matured concrete. He described the action he took: "So we put in six two-and-a-half-inch-diameter steel cables all around the pedestal and tightened them to two hundred fifty thousand pounds of tension each. To me that was the most dangerous part of the whole job. Can you imagine how much energy is hidden in there? Those cables are still there, covered with concrete. I told them to put up a notice just in case someone goes in there with a core drill someday" (Wells 1999).

While no one thought it would be easy, no one expected the "bad" concrete to be so incredibly difficult to break up. In the first attempt, one-inch holes were drilled a foot apart and six feet deep and filled with a proprietary expansion agent. Although some few cracks appeared after twenty-four hours, the demolition contractor was quite unable to break the concrete with his air-powered jackhammers. When they brought in huge hydraulically powered machines, the pedestal shook so much that Wells stopped the work while he figured out a better way to fracture the concrete. Finally a more powerful expansion agent was found and, by cracking off the outer layer of concrete first and cutting away the dense labyrinth of embedded one-inch reinforcing bars, the workmen were at last able to make progress, albeit at a slow pace. It took a demolition crew of four men about five days to remove a segment of concrete forty feet long. When the job was finished, one of them remarked, "If this is bad concrete, I hope I never have to work on removing good concrete." Figure 9.2 shows the concrete removal work in progress.

Using the new expansion agent, removal of the old concrete and its replacement with new concrete proceeded in a well-coordinated sequence.[4] As soon as

Figure 9.2. Removing Concrete from Goldstone Antenna Pedestal, July 1983. A section of the outer layer of concrete has been removed from the haunch to expose the steel reinforcing bars. In the center of the picture, a portion of the temporary antenna support structure passes upward through the work platform, to an anchor point at one corner of the base triangle. The siding above the haunch encloses electrical and hydraulic machinery space in the base of the alidade. The presence of workmen on the platform conveys the immense size of the pedestal base. NASA/JPL.

each set of three sections was cleared of concrete and reinforcing steel bar, new rebar was welded into position and new concrete was poured and allowed to cure for a week before work started on a new section. Great care had been taken in selecting the source of aggregate that was to be used for the new concrete mix. Samples of material from several quarries were tested for alkali content. Unfortunately, the only aggregate suitable for the purpose was located in a quarry near Mountain View, California, about 400 miles from the Goldstone site. Nevertheless, 3,000 cubic yards of this hard limestone rock were trucked to Goldstone, mixed in batches, and pumped into place on the haunch. The concrete in the first batches was so stiff that the big pumpers were unable to move it. The specified proportions of aggregate, cement, and water had been followed exactly. More water would weaken the mix. What could be the problem? Wells finally decided that the limestone aggregate was itself absorbing water from the mix and that the water content could be adjusted without detriment to the ultimate strength.

And so it proved to be. After the concrete had "cured" properly, core samples were taken from each of the batches around the new haunch and tested. Not one of them failed to pass the strength test.[5] Displaying his typical pragmatic approach to problem solving, Wells (1999) recalled the aggregate water absorption problem: "Some people were skeptical about the aggregate absorbing water, so I took some aggregate and put it in a quart jar with half a cup of water. Half an hour later the water was gone. They were convinced after that." Wells's test showed that the aggregate did indeed absorb water from the concrete mix, as he had thought, and convinced the skeptics.

When the sole plates, runner, and pads were reinstalled and aligned, the 64-meter antenna was transferred back to the hydrostatic bearing, tested, and returned to service one day ahead of schedule.

In his final report McClure (1999b) observed, "Raising the six-million-pound antenna and placing it on a temporary structure, removing the hydrostatic bearing, and removing and replacing 450 cubic yards of concrete was a monumental task. The task was completed on schedule, within budget, and without a lost-time accident. The concrete required working with state-of-the-art technology and the development of new design considerations for high MOD E concrete.[6] As a result of this work, the replacement concrete now provides a stiff ring around the pedestal which will provide excellent support for the hydrostatic bearing at DSS 14 for many years."

As McClure's on-site engineer, Dale Wells had played a major part in bringing the task to its successful conclusion. The citation to the Exceptional Service Medal presented by NASA for his work says it best: "For his exceptional ingenuity and engineering skill demonstrated in the very successful completion of the repair of the hydrostatic bearing and concrete pedestal of the Goldstone antenna. This was the most difficult repair ever undertaken by the Deep Space Network." As McClure had so succinctly observed, it was, in the truest sense of the word, a "monumental" task.

Problems with Procurement

With the Goldstone pedestal repair work progressing well under Wells's vigilant supervision, McClure and McLaughlin were able to turn their attention to procurement contracts for critical components of the antenna upgrade part of the overall project. Knowing that the lead time for fabrication of the three cast aluminum subreflectors would be the longest, McClure decided to start the time-consuming procurement process immediately.

The subreflector was a large but very delicate piece of solid precision-shaped metal. Twenty-six feet in diameter, about six feet deep, and five-eighths of an

Figure 9.3. Subreflector and Surface Shaping Machine. The two half-castings were bolted together for the surface shaping process that was carried out by this special digitally controlled three-axis milling machine. One of only three in the world, this machine in Santander, Spain, produced a precision surface contour whose measured accuracy, 0.006 inches, considerably exceeded the specified value of 0.010 inches. NASA/JPL.

inch thick, it was probably the largest cast aluminum structure of its type that had ever been built, and it required state-of-the-art casting, welding, and machining techniques for its construction. Unlike a surface of revolution that could be turned on a regular spinning lathe, the surface was a unique shape that required the use of a very large, digitally controlled three-axis milling machine to produce the desired accuracy of 0.010 inches.

In April 1985, JPL awarded a contract for the production of one subreflector, with an option for two additional units, to a large engineering company in California. This firm subcontracted the actual fabrication work to a Spanish general engineering company, which in turn subcontracted the work to three small Spanish companies that specialized in casting, welding, and machining respectively. All three fabrication processes proved to be extraordinarily difficult and incurred substantial cost overruns and delays. The final product, however, was of superior quality. It fully met, and in some instances exceeded, the design specifications for the surface contour. Although the quality of the finished product was very satisfactory, McClure was greatly concerned about the late delivery. The complexity of the milling machinery required to produce the subreflector is seen in figure 9.3.

Because of its size and fragility, it was originally intended to disassemble the completed subreflector and ship it to the antenna site in two pieces. This added complexity was avoided by moving it in a single piece, fortunately without any deleterious effects. A template check made at the site after arrival showed no change in the shape of the precision surface. This solved the schedule problem for the first subreflector, but it did not solve the cost overrun problem.

In February 1986, JPL exercised its contractual option and authorized procurement of the castings for the second and third units. In the weeks that followed, McClure constantly pressed the California contractor for a report on progress in Spain. But despite his concern about the schedule and cost of the two additional units, it was not until September that he learned that the Spanish subcontractors refused to begin work until the price was adjusted upwards. To avoid a catastrophic impact on his downtime schedule, McClure had no option but to meet their demands. Finally production of the second and third units got under way in Spain, but the unrecoverable time lost would have a ripple effect through the entire project.

McClure now turned to the next item on his priority list, the new high-precision metal panels that would form the specially shaped reflector surfaces for the 70-meter antennas at all three sites. This second procurement contract that JPL put out for bids covered a set of reflector panels for the Madrid antenna, with an option for two additional sets, for Canberra and for Goldstone. After due process the prime contract again went to a California company, the same one that had won the contract for the subreflector. And as before, the task was subcontracted to an overseas firm, this one in Bergamo, Italy, a specialist in aluminum construction. The contract was for 1,289 panels of precision-shaped and bonded-aluminum construction, with a surface accuracy of 0.125 mm (0.005 inches).[7] The Bergamo firm constructed a special facility for the purpose that, at the time, contained one of the most advanced aluminum cleaning and bonding plants in Europe. The result was a superb set of panels that met the specifications in every way, including delivery time. This very satisfactory result was not, however, achieved without some cost. During the course of the contract, the U.S. dollar suffered a dramatic decline in value, to the point where the Italian contractor stood to suffer a serious loss on his dollar-based subcontract. Since neither the U.S. prime contractor nor the Italian subcontractor could be expected to absorb the entire loss, McClure and McLaughlin were again forced to negotiate an equitable solution. JPL funded about half the loss, while the other parties absorbed the remainder between them.

Later McClure (1999b) observed: "These two situations showed the difficulty of working with a second-level subcontractor, a difficulty compounded by the fact that the subcontractor was located overseas, separated not only by the miles but by language and cultural differences as well."

In accordance with the terms of the international agreements between the United States and Spain and Australia, the fabrication of rib trusses, center-section reinforcement, quadripod, and counterweights, as well as the assembly and actual erection work, was carried out by local contractors in their respective countries. NASA transferred funds to the appropriate authorities in Spain and Australia to cover the costs involved in carrying out the work. The work was executed expeditiously, and without significant complications attributable to the contractors. Steel fabricators and constructors from Long Beach and Oakland, California, performed the corresponding work for the Goldstone antenna.

Augmentation of the Apertures

Madrid

Three months after it received the contract, a Spanish construction company began erecting a labyrinth of steel scaffolding that would give its steelworkers access to the outer rib trusses high on the perimeter of the existing 64-meter antenna when it was stowed in the zenith (vertically-pointing) position. Those trusses would eventually be replaced by larger, specially shaped ribs that would support the contoured surface of the new 70-meter antenna.

It was midsummer in Spain. At the antenna site at Robledo the weather was calm and clear, sunny and hot, perfect for high-steel construction work, and the Spanish workers made rapid progress. Carefully situated around the antenna were the new rib trusses, the steel beams for construction and reinforcing, the counterweights, and the quadripod, all standing ready for placement on the central hub as soon as the disassembly work on the old 64-meter antenna was completed. There they would remain until McClure got approval to take the 64-meter antenna out of service for the upgrade work.

McClure was very much aware of the domino effect that a false start in Spain would have on the timely execution of the upgrades in Australia and Goldstone. Any delay would ultimately impact the in-flight requirements of *Voyager 2*, then on its way to Neptune, of *Galileo* and *Ulysses*, about to be launched for missions to Jupiter and solar orbit respectively, and of a fourth spacecraft, later known as *Magellan*, that was being prepared for a mapping mission to Venus. All four were major NASA planetary missions and, once they were launched, all required the

70-meter antennas at specific, unalterable times. McClure's scheduled start and finish times simply had to be met.

It was standard NASA/JPL management practice to establish an independent board of review to assess overall readiness to initiate a nonrestartable course of action—a spacecraft launch, a flyby, entry into a planetary orbit, the implementation of new DSN capability. The 70-meter upgrade project fell into this last category, and McClure established a Downtime Readiness Review Board to review the status of the project and assess its readiness to start the Madrid downtime, the event that would initialize all of the subsequent global action. It convened at the Madrid site for three days in early July 1986.

After hearing the technical presentations, the board was satisfied that the established criteria had been met. Although the subreflector had not been completed, the prospects for timely delivery were good. The board gave its approval to start the downtime as planned, and McClure gave the go-ahead.

Sometime in the early hours of August 1, 1986, right after the last scheduled tracking pass, a technician at the Madrid station pointed the 64-meter antenna to its zenith position, put it in "park" mode, and shut it down. After thirteen years of continuous service, the Madrid 64-meter antenna was about to pass into DSN history.

At daybreak the construction workers arrived on-site and began the first task in McClure's erection plan, disassembling the 64-meter reflector. In the prevailing fine dry summer weather the work progressed rapidly as, module by module, large sections of rib assemblies with panels attached were unbolted or cut free of the inner rib structure and set aside to make way for the new assemblies. Within a couple of weeks the original antenna had been reduced to the center section and quadripod. At that point the first sign of trouble appeared.

It began with a major mechanical problem on the 500-ton crane, which delayed the completion of the disassembly work. When that was corrected and the buildup began, it was found that the new quadripod attachment points would not fit the existing center hub structure. By mid-September, much to McClure's dismay, McLaughlin estimated that the project was already three weeks behind schedule.

Bad as that was, there was worse to come. Delays mounted when steelworkers discovered that misaligned connection points on the new rib trusses required rework to make them fit the former structure. Efforts to recover the lost time were futile.[8] On October 1, 1986, the Spanish contractor advised McClure's on-site manager, Otto Rotach, that the downtime would have to be extended by eight weeks.

Suddenly McClure found himself between a rock and a hard place. The upgrade work on the three antennas was to be carried out in serial fashion. He

Figure 9.4. Outer Truss Modules Installed on New 70-Meter Antenna. At this stage, about half of the outer truss modules have been attached to the completed ring of inner truss modules. The quadripod and elevation wheel with counterweight are clearly visible. The huge scaffolding to the left provided working access to the underside of the rib trusses. The tower of the large crane that lifted the trusses into place runs vertically through the center of the picture. NASA/JPL.

could not delay the start of the Australian and Goldstone downtimes, yet he could not start the Australian work on schedule without compromising the basic NASA guideline of never having two antennas down simultaneously. What was to be done? As we saw earlier, the answer came from a totally unexpected direction.

In the aftermath of the tragic accident with space shuttle *Challenger*, NASA delayed the *Galileo* mission to Jupiter until an alternative upper-stage launch vehicle could be found. With *Galileo* no longer a constraining factor on the antenna downtime, JPL elected to suspend its two-antenna rule and permit McClure to start the Australian downtime as originally planned. Since the *Voy-*

ager 2 Neptune encounter then became the driver for completion of the project, the downtimes for both Australia and Goldstone were lengthened to eight months to provide an additional margin.

Meanwhile at Madrid, work on the new antenna proceeded apace—behind schedule, to be sure, but without significant further delay. The photograph in figure 9.4 shows how the antenna looked at this stage of the erection process.

By mid-December the surface panels, all 1,272 of them, had been installed and the delicate optical alignment phase began. Ideally, that type of work required clear, still, dry weather for the precision theodolite sighting and optimum adjustment of individual panels. That, however, was not to be. As fall passed into winter, the weather at Madrid turned bad. By some estimates, the winter of 1987 brought the worst weather Europe had seen in a century. Delayed by rain, hail, and high winds at the site, the optical panel alignment took twice as long as originally planned and was not completed until the end of January. Despite the inclement weather, the engineers achieved an overall accuracy figure for the new 70-meter surface of 0.55 mm (0.022 inches) which, while leaving plenty of room for improvement, was within the specified limits.

In the original construction sequence, the installation of the subreflector and counterweights was to follow the alignment of the panels and lead to the final stage, the servo tests. Because delivery of the subreflector was delayed, this sequence could not be followed. Instead, a large concrete block of the same weight as the subreflector was installed as a substitute during the weight and balance tests. By the time the subreflector arrived on-site in early March and was installed, the counterweights had also been installed and the servo tests had been satisfactorily completed.

With the subreflector installed and aligned, the final step involved a measurement of the all-important RF gain of the new 70-meter reflector. According to the review board's criteria for returning the antenna to service, the RF gain had to equal or exceed that of the 64-meter antenna as measured just prior to the downtime. Early in April 1987, while McClure figuratively held his breath, Spanish antenna engineers ran the test on the new 70-meter antenna in exactly the same way that they had run the test on the 64-meter reflector eight months earlier.

Back at JPL, McClure got the news by phone from Madrid very early on a blustery Southern California spring morning. The new antenna satisfied the basic criteria but was behaving in a distinctly "anomalous" way as it moved through its full range of elevation angles. The cause for concern may be seen in figure 9.5.

McClure called Bathker to "share the pain" and decide what to do next. Fully confident of the efficacy of the basic design, Bathker suggested that the root of

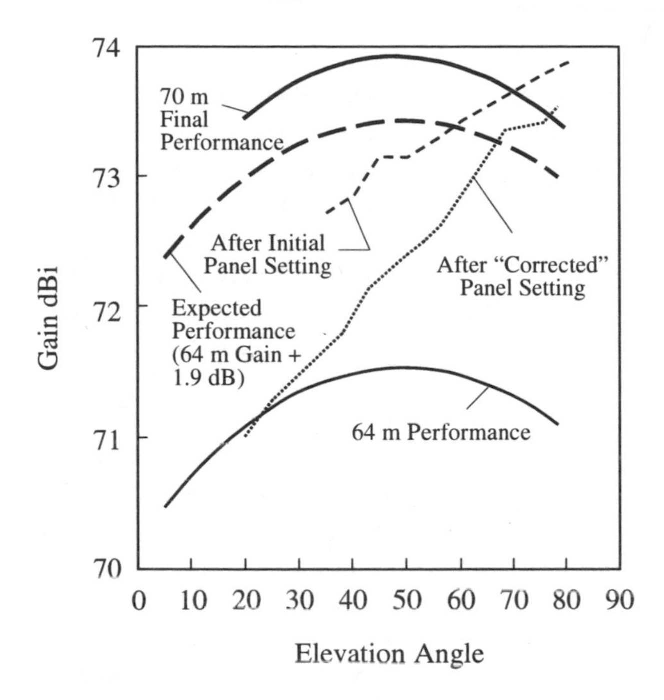

Figure 9.5. Gain versus Elevation Angle for Madrid 70-Meter Antenna. The graph shows how the RF gain changed as the antenna moved through its full range of elevation angle. The lower curve displays the measured 64-meter performance, while the broken upper curve (May 1988) displays the expected 70-meter performance. Between the two, the dotted curves show the results with initial and corrected settings. The discrepancies caused great concern in April 1987. Illustration by Fred D. McLaughlin, NASA/JPL.

the problem with the gain lay in the settings of the reflector panels. But both he and McClure were assured that the optical settings had been made correctly despite the inclement weather at Madrid. At a loss to identify an obvious source of error and under considerable pressure to resolve the problem quickly, McClure and Bathker ordered an immediate review of the computer software that converted the optical theodolite readings for each panel to an appropriate mechanical adjustment that could be made by a technician at the reflector surface. A quick scan suggested that a program error could have resulted in a reversal of the mechanical adjustment instructions. It appeared that all four adjustments on all 1,272 panels had been made in the wrong direction—"in" rather than "out," "up" instead of "down," and vice versa. With time running out and no better option to hand, McClure ordered the technicians at Madrid to go back onto the antenna surface and readjust all the panels to "twice the original value in the opposite direction." That completed, they were to repeat the gain measurement.

The outcome of that effort is shown in figure 9.5. Rather than improving the antenna's performance, the "corrected" panel settings had made it worse.

This was a grave situation that called for immediate action. McClure (1999b) described his response: "At this point in time, panel setting on the Australian antenna was scheduled to begin in six weeks, and the Goldstone antenna down-time was scheduled to begin in three months. On May 7, 1987, a 70-meter Antenna Anomalous Performance Tiger Team[9] was formed, and an intensive effort was mounted to understand the nature of the problem and to develop a solution. In the meantime, since the performance was at least as high as the 64-meter antenna's had been, a decision was made to return the antenna to operations while the Tiger Team investigated the problem."

DSS-63 Madrid resumed operational tracking support on May 14, 1987, after nine and a half months of frenzied conversion and upgrade activity. Now 70 meters in diameter, it had a higher gain than the 64-meter antenna it replaced, but only just. Still, the changes in priority for NASA's planetary missions that resulted from the *Challenger* accident now provided McClure with more flexibility in dealing with unexpected contingencies. As he saw it, the 70-meter antenna's upgraded performance was of lesser importance than its availability for tracking support. Later there would be time for refinements.

Over the ensuing months, a number of structural and other improvements were made to the Madrid 70-meter antenna in accordance with the Tiger Team's recommendations. At the first opportunity, the panels were reset yet again, this time using a high-precision holographic method. Although it required the antenna to be out of service for several days, the results proved to be well worth the trouble. After remeasuring the performance, Madrid antenna engineers were pleased to report that the RF gain not only met the specified value—1.9 dB greater than the 64-meter antenna—but also exceeded that value by a handsome margin at all elevation angles. Furthermore, the measurements showed that the angular shape of the antenna's radio beam was in good agreement with the shape predicted from the theoretical design. Bathker was very pleased, and McClure breathed a deep sigh of relief.

Final adjustments to the gain performance were hampered by poor weather at Madrid and by difficulty in getting time on the heavily committed 70-meter antenna to make further holographic measurements, so it was well into 1988 before a value for the ultimate RF gain could be established. The official value exceeded the RF gain of the 64-meter antenna by 2.1 dB (a factor of 1.6), a result that clearly vindicated Bob Stevens's original faith in the enterprise, and that was to prove of inestimable value to NASA's deep space program in the years to come.

Figure 9.6. DSN's 70-Meter Antenna at Robledo near Madrid, Spain, Completed in May 1987. NASA/JPL.

The photograph (figure 9.6), taken shortly after completion of the upgrade in May 1987, affords some appreciation of the enormous size and complexity of the rigid backup structure that supported the main reflector surface.

Canberra

McClure's plans for upgrading the 64-meter antenna in Australia took a slightly different path from the plan used in Spain, and a rather more complex one. JPL intended to take advantage of its multiple procurement options to provide the new subreflector, panels, and positioner for the Australian 70-meter antenna. It also transferred funds to the Australian government to cover a construction contract with a local company for the fabrication of the new rib trusses, reinforcing steel, and quadripod and for the disassembly and erection work. The Australian contractors completed all the preliminary fabrication and delivery on schedule, and in mid-January 1987 McClure once again convened the Downtime Readiness Review Board, this time at the Canberra site. Using the same criteria as for the Madrid antenna, the board approved the start of downtime for February 1, 1987, as planned, but expressed grave concern about use of an "interim" subreflector in place of the actual subreflector, which had not yet been delivered by the Spanish manufacturer.

It will be recalled that the Spanish contractor for the subreflectors had experienced serious manufacturing problems with the first subreflector, necessitating changes to subsequent deliveries. Almost a year before that happened, however, McLaughlin and McClure had drawn up a contingency plan to construct a fiberglass subreflector of the same weight and dimensions, though not the same shape, as the real cast-aluminum subreflector. This substitute device would be a simple surface of revolution and therefore relatively simple and cheap to fabricate. The surface would be covered with expanded aluminum mesh, several panels of which had been tested by Bathker's people at JPL and found acceptable for the purpose.

In November 1986, when the first shadow of doubt began to appear about the Spanish delivery schedule, McLaughlin urged McClure to put the contingency plan into effect. A contract for a fiberglass subreflector was immediately placed with a California plastics company. Three months later, just after the Australian downtime started, the "interim" subreflector was shipped to Sydney in two pieces and trucked to the Canberra site, where it was subsequently installed on the new 70-meter antenna structure without incident.

Although the erection work took somewhat longer at Canberra than at Madrid, it proceeded smoothly, did not require any extensive on-site rework, and was not subject to significant delays due to bad weather. In fact, the work proceeded so expeditiously that when the time came to align the panels on the Canberra reflector, the Tiger Team at JPL had not yet determined the cause of the anomalous performance of the Madrid antenna.

McClure could not wait for a resolution. He instructed the Canberra antenna engineers to begin by setting the panels using the original procedure, since there was no credible evidence to support the incorrect-software theory, and then to use the precision holographic technique for the final setting.

Although this decision added three weeks to the downtime schedule, it was time well spent. The final measured gain fell only a little short of the desired 1.9 dB improvement over the 64-meter gain. Bathker attributed most of the degra-

Figure 9.7. DSN's 70-Meter Antenna at Tidbinbilla near Canberra, Australia, Completed in October 1987. NASA/JPL.

dation to the temporary use of the symmetrical "interim" subreflector. DSS-43 returned to operational service in the new 70-meter configuration on October 21 1987, a little less than nine months after the upgrade work began.[10]

Eventually the Spanish contractor completed the third and final precision-shaped subreflector. It was shipped directly to Canberra and mounted on the antenna in May 1988. The final holographic setting was completed a month later, when the ultimate RF gain was determined to be 2.16 dB relative to the 64-meter performance, a figure that met the project goal and stood slightly above the measured performance of the Madrid antenna. Although the precision shapes of the subreflector and main reflector surface are not discernible in the photograph (figure 9.7) of the Canberra antenna, the arrangement of the tricone surmounted by the three microwave feed horns that are illuminated by the subreflector is clearly seen.

Goldstone

By comparison with the fabrication and erection work at the overseas sites, the work for the 70-meter antenna at Goldstone went forward with little or no perturbation to the original plan and schedule. McClure's 70-meter project team at JPL had acquired a lot of firsthand experience by then, and it was much easier to correct management and supply problems, since most of the project was executed within California.

JPL chose its contractors well. A steel construction company from Long Beach delivered the new rib trusses, center section reinforcing material, counterweights, and quadripod to Goldstone on time and well before McClure planned to start the downtime. Almost immediately the Oakland construction company charged with the disassembly and erection work started moving equipment on-site in preparation for a start as soon as the 64-meter antenna became available. As he had done in Spain and Australia, McClure convened the Downtime Readiness Review Board to assess the situation and approve a start on the upgrade work.

The board's deliberations were brief. After a few quibbles about potential conflicts with other work at Goldstone and about a contingency plan to procure yet another "plastic" subreflector "just in case," the board gave the go-ahead. The final stage of the 70-meter upgrade project began on October 1, 1987.

This time there were no glitches. The precision-shaped subreflector arrived at Goldstone in February 1988, a few weeks ahead of the time it was needed in the erection sequence.

Just before the subreflector was installed, however, there occurred a curious incident that emphasized the importance of maintaining constant vigilance in carrying out even the apparently simplest tasks on a structure so huge and yet so deceptively fragile. McClure (1999a) recalled:

> One day at JPL, my phone rang and one of the engineers from Goldstone said, "Guess what? They've dropped a sledgehammer through the new feed cone of the seventy-meter antenna, and smashed the S-band microwave horn." It seemed impossible, because the horn was mounted on the top of the tricone, over a hundred and fifty feet above ground and protected at all times with a cover. Nevertheless it was true. An ironworker standing at the apex of the quadripod had been pounding on the structure with an eight-pound sledgehammer when the head broke off and fell directly onto the feed horn fifty feet below. Although the mouth of the horn was protected by a thick sheet of plywood, the hammerhead passed right through it and totally destroyed the precision microwave device.

It was an expensive accident for the erection contractor, whose insurance company paid JPL $300,000 to replace it. Fortunately a spare unit was available, and the accident did not result in any significant delay to the on-site work.

Because the JPL Tiger Team had not resolved the problem with the anomalous performance of the Madrid antenna by the time McClure was ready to begin the reflector panel setting sequence, he and Bathker decided to go ahead with the same techniques that had proved successful on the overseas antennas. Highly confident in the microwave holography techniques used in Spain and Australia, they spent less time on the preliminary optical alignment and relied entirely on the holographic measurements to achieve the final precision setting of the reflector surface.

The final words belonged to Don McClure. Reporting the last scheduled activity for the project, he wrote: "As a result of the experience gained in Spain and Australia the final panel settings and antenna efficiency measurements were conducted in a smooth manner enabling the Goldstone antenna to be returned to operations, with 70-meter performance, on May 29, 1988, two days ahead of schedule" (McClure and McLaughlin 2001).

The aerial view of the Goldstone 70-meter antenna in figure 9.8 gives a good impression of its size by comparison with the nearby buildings and vehicles.

It is no exaggeration to record that McClure's final report was received at JPL and NASA headquarters with great satisfaction and much relief. The attempt to make the Big Dishes even bigger had always been regarded in some quarters as unduly risky. The 64-meter antennas were at once irreplaceable and essential to NASA's then highly visible planetary missions. And all of the in-flight missions

Figure 9.8. DSN's 70-Meter Antenna at Mars Site, Goldstone, California, Completed in May 1988. The photograph emphasizes the enormous size of the main reflector surface by comparison with nearby buildings and vehicles. NASA/JPL.

were inconvenienced by the continuing antenna downtimes. Those that did not require the support of 64-meter antennas were required to share observing time on the smaller 34-meter antennas with those that did.

With just over a year to go before it reached Neptune, *Voyager 2* would depend completely on the new 70-meter antennas, along with an elaborate array of other antennas, to capture its imaging and other science data from unimaginable depths of space. With the passage of time, the repercussions of the *Challenger* accident had settled out and the planetary missions most directly affected—*Magellan, Galileo,* and *Ulysses*—had been replanned and were waiting in the wings for the resumption of space shuttle launches in 1989 and 1990. All three missions now required 70-meter antenna performance for successful accomplishment of their primary mission objectives.

At NASA headquarters, initiatives for new planetary missions had been quick to capitalize on the potential for even greater science returns from deep space than were afforded by the DSN's enhanced tracking capability. The same ideas had occurred to the planetary radio astronomers and the planetary radar scientists, who quickly oversubscribed the limited amount of 70-meter antenna time that was made available for those purposes.

A factor of 1.6, which roughly represented the project's goal of 1.9 dB improvement over the link performance of the 64-meter antennas, would not, in many other contexts, seem a really significant step forward. But the incredible technical difficulty of realizing any improvement in the figure of merit of large antennas operating at that level of performance made the success of the 70-meter upgrade project a major achievement indeed.[11]

Jewels in the Crown

If the Big Dish in its new configuration was a crowning achievement for the DSN, the low-noise amplifiers hidden deep within the tricone were, quite literally, the jewels in the crown. Over the intervening twenty years since the 64-meter version of the Big Dish had entered service with the DSN, the masers had been substantially improved in both maintainability and performance. Variations in the gain of the maser that were associated with relative movement of magnet and maser due to the tipping motion of the antenna had been minimized by replacing the massive permanent magnets with lighter, compact superconducting magnet assemblies that allowed the maser assembly to be placed within the magnet coil itself. Combined with advancements in the microwave traveling-wave structure of the internal ruby crystal, these changes resulted in much improved values for gain, stability, bandwidth, and noise figure.

Paralleling the S-band maser work, the maser group at JPL incorporated these improvements in the early X-band maser designs also. Following a period of development and evaluation in the early 1970s, X-band downlinks finally came into full operational service in the DSN in 1975, supported by robust, dependable X-band masers. The 1980s saw further refinement of the S-band and X-band masers to meet the requirements of ever more ambitious planetary missions, most notably the Voyager encounters with Saturn in 1981 and Uranus in 1986. As a direct consequence of this work, the new 70-meter antenna entered operational service in 1987 with four highly developed and thoroughly field-proven models of these exquisite devices, two for S-band, and two for X-band.

These were the jewels that, together with the 70-meter antenna itself and an ensemble of high-powered S-band and X-band transmitters, formed NASA's awesome instrument for unmanned exploration of the solar system that, in time, came to be known as the Big Dish. These were the operational systems that the DSN used on a routine basis to maintain continuous uplink and downlink communications with dozens of robot spacecraft that were engaged in scientific missions to the planets and many other parts of the solar system. But that was not all.

The DSN's Big Dish served two other functions that did not require a spacecraft for their science experiments. They were radio astronomy and planetary radar (Mudgway 2001).

In radio astronomy experiments, the tremendous collecting power of the Big Dishes was used to observe the emanations from extraterrestrial radio sources, like quasars and radio stars. Analysis of data from these experiments contributed to the body of scientific knowledge regarding the distribution, origins, dynamics, structure, and composition of distant parts of the universe. Radio astronomy made use of extraterrestrial sources of radiation, and depended for its success upon the sensitivity of the Earth-based receiving system.

Planetary radar, on the other hand, depended upon the strength of the Earth-based transmitting systems that it used to "illuminate" its various target bodies of interest—planets, comets, and eventually Earth-crossing asteroids—with S-band or X-band radio energy. Interpretation of the radio echoes reflected from the surface of a target body yielded a wealth of scientific information about its shape and rotation, surface features, composition and roughness, and reflectivity. When used for planetary radar studies, the Big Dish was called the Goldstone Solar System Radar.

Both of these enterprises required special receiving and high-power transmitting equipment, which resided in the third cone of the tricone structure. In addition to their intrinsic scientific value, experiments such as these had been of great benefit to the DSN, because much of the resulting research-oriented technology was eventually used to enhance the capabilities of the Deep Space Network. The masers, microwave components, and high-power transmitters that now formed an integral part of the new 70-meter antenna facility were important examples of this interaction.

Typical values for a selection of common telecommunications parameters for the DSN 70-meter antennas are shown in table 9.1.

NASA rightly recognized the 70-meter antenna upgrade as a significant achievement with Group Awards to the engineers and others who had been involved the project. In appreciation for the 1.9 dB of telecommunications link improvement that Don McClure's timely completion of the upgrade provided

Table 9.1. Nominal Values of Selected Telecommunications Parameters for the DSN 70-Meter Antennas

Parameters	S-band	X-band
RF receive gain	63 dB	74 dB
Recommended minimum operating carrier-signal power	–170 dBm	–165 dBm
Operating system noise temperature*	19 K	21 K
with ambient diplexer	21 K	
with cooled diplexer		21 K
Antenna figure of merit (G/T)	50 dB	61 dB
RF transmit gain	63 dB	73 dB
Transmit power for		
planetary spacecraft	20 kW, 400 kW	20 kW**
planetary radar Goldstone	400 kW	400 kW

*These are theoretical or "in vacuum" numbers. In a normal application for a link budget design, they would be adjusted for the atmospheric and environmental losses associated with local weather and with the changes in elevation that occur during the antenna's eight-hour view period of a planetary spacecraft. Error margins would also be included in the link budget.
**X-band transmit capability was added in 1998.

for *Voyager 2*, NASA awarded him its Exceptional Engineering Achievement Medal. The citation read: "In recognition of outstanding achievements in the design and implementation of the 70-meter subnetwork of the Deep Space Network, which permitted a more complete examination of Neptune and greater scientific value from the Voyager Encounter." McClure would have been the first to acknowledge that, while there could be only one medal, the honor was properly shared by all those who worked with him to earn it.

In the rush of events that followed completion of the Goldstone 70-meter antenna, the tremendous new network capability soon became a matter of fact, and was quickly absorbed into the regular routine of the network's deep space operations. Over the next fifteen years, the three Big Dishes contributed in a major way to NASA's most ambitious and far-reaching missions to the edge of Earth's solar system, and demonstrated a capability to reach even beyond that uncertain boundary, into the uncharted regions beyond.

10 Closing Scenes

From the Past

Voyager's Grand Finale, August 1989

The DSN had changed significantly in the five years that followed Bob Stevens's original proposal for the upgrade of the 64-meter antennas. Most apparent of these changes was, of course, the increased diameter of the large antennas. This, together with the vastly improved microwave optics of the new reflector surfaces, increased their ability to receive weak signals from distant spacecraft by about 2.1 dB (a factor of 1.6). Goldstone, the last of the three to be completed, had come into service in May 1988, and now, a year later, all three 70-meter antennas were about to demonstrate their impressive new capabilities by capturing the science and imaging data from *Voyager 2* as it made its long-awaited encounter with Neptune.

Still, impressive as the new antennas were, they could not by themselves recover science data from Neptune of the same quality and quantity as those from Uranus. To do that, it would be necessary not only to couple the 70-meter antennas with all of the smaller 34-meter antennas at each complex but also, at Goldstone and Canberra, to augment the NASA arrays with the largest non-NASA receiving antennas in the United States and Australia respectively (Stevens

1984). The effective increase in size of the combined receiving antenna aperture would compensate for the tremendous increase in radio transmission distance between Earth and Neptune (30 AU) over that between Earth and Uranus (20 AU). In short, to cover the *Voyager 2* encounter with Neptune, the DSN planned to implement the largest assembly of microwave receiving capability anywhere in the world.

In the intervening years, while Don McClure and his team were involved with the onerous task of upgrading the 64-meter antennas to 70-meter diameter, other engineers at JPL, notably Jim Layland and Don Brown, had been working with Australian radio astronomers at Parkes and with American radio astronomers at Socorro to perfect the complicated arraying techniques that would allow the antennas of those agencies to effectively augment the NASA antennas (Layland and Brown 1985; Layland, Napier, and Thompson 1985). An aerial view of the Very Large Array (VLA) at Socorro is shown in figure 10.1, while figure 10.2 shows the 64-meter radio telescope at the Parkes Observatory in New South Wales, Australia.

Most of the specialized equipment required for arraying the Parkes antenna with the DSN antennas at the Canberra complex had been developed for the Uranus encounter in 1986. It was already in place and needed only minor modification to accommodate the larger 70-meter antenna and the additional 34-meter antenna that Stoller's people had added to the complex since then. However, arraying the twenty-seven 25-meter antennas of the VLA with the Goldstone 70-meter antenna, some 600 miles distant, presented an entirely different set of technical problems, all of them challenging and all of them expensive. Formidable as these difficulties appeared at the start, they were all eventually overcome, so that by mid-1989 when *Voyager 2* began to approach Neptune, the DSN arraying configurations with both Parkes and Socorro were completed, fully tested, and ready to cover the spacecraft's science observations during its close flyby of the planet and its satellites beginning in August (Mudgway 2001).

It was a mind-boggling arrangement, impossible for anyone to observe in its entirety, difficult to comprehend in its complexity. It was a consummate example of innovative deep space communications technology, not seen in the past and not likely to be replicated in the near future. At the outer reaches of the solar system *Voyager 2*, a tiny spacecraft that had been operating continuously in the hostile environment of deep space since its launch twelve years earlier, was about to stage the grand finale to a mission that had taken it, in succession, to spectacular encounters with Jupiter in 1979, Saturn in 1981, and Uranus in 1986. Now in mid-1989, three billion miles distant, moving under the influence of the dynamical and gravitational forces that had determined its motion since

Figure 10.1. Very Large Array (VLA) at Socorro, New Mexico, 1988. Situated about fifty miles west of Socorro, the VLA radio telescope is owned and operated by the National Radio Astronomy Observatory (NRAO). The VLA is formed by twenty-seven 25-meter parabolic antennas, positioned along a Y-shaped rail track as required by individual radio astronomy observations. In a cooperative effort with the DSN, each antenna was fitted with X-band receiving equipment to support the 1989 *Voyager 2* encounter with Neptune as part of a Goldstone/VLA array. The significant science data return that resulted from this complex arrangement included images of the planet and its satellite Triton, and enabled the detection of rings around Neptune. Subsequently the Goldstone/VLA array was used to generate the first planetary radar images of Saturn's satellite Titan. NASA/JPL/NRAO.

leaving Earth, *Voyager 2* moved inexorably to intercept the path of the huge, gaseous planet Neptune. Controllers at JPL had adjusted its trajectory so precisely that it would cross the path of Neptune just far enough away to avoid capture by Neptune's gravity but close enough to afford its cameras and other scientific instruments humankind's first close-up view of the mysterious blue planet and several of its satellites. What would they see? What new science would they find?

The answers to such questions were embodied in the data stream sent by the spacecraft's radio transmitter, with the power of a 25-watt lightbulb, to the incredibly sensitive antennas of the Deep Space Network. Clustered at three evenly spaced locations around the globe and electrically anchored to the mighty 70-meter antenna at each location, the antennas and receivers of the DSN, CSIRO, and NRAO identified *Voyager 2*'s ephemeral radio signal amid the chaos of radio and cosmic noise in which it was embedded, and instantly captured it electronically to ensure that it did not slip away with the rotation of Earth and the detuning effect of the Doppler shift between Earth and Neptune. That done, there began the complex process of combining the signals from each of the separate antennas to produce an enhanced composite output from which the precious science data could be extracted. Further steps in the electronic processing path involved demodulation and decoding, each a complex technique in

Figure 10.2. Parkes Radio Telescope, 1988. Located 220 miles west of Sydney, Australia, the Parkes Observatory is part of the Australia Telescope National Facility, a division of the Commonwealth Scientific and Industrial Research Organisation (CSIRO). The 64-meter antenna was built by the CSIRO division of radio physics in 1961, and continual upgrades of the reflector surface, computers, and electronics have kept it in state-of-the-art condition. This antenna contributed in no small measure to the design of the DSN 64-meter antennas, and has always maintained a close and productive working relationship with the DSN's Deep Space Communications Complex near Canberra, some 120 miles to the south. The rural township of Parkes, population 9,500, lies in a rich mining and sheep- and wheat-producing area. NASA/JPL/CSIRO.

itself, after which the data, now in digital form, was transferred via satellite and microwave communications links from each complex to anxiously waiting scientists, engineers, and navigators at JPL in Pasadena.

By arrangement with CSIRO and NRAO, the DSN began tracking *Voyager 2* in full array configuration well before its closest approach to Neptune. These agencies had graciously agreed to make their antennas available for arraying with the DSN antennas from about six weeks prior to the closest approach until about four weeks after the event. Although their scientists were intensely interested in the results of the Voyager encounter, they had to pursue projects of their own and meet schedules that had been suspended to make way for Voyager during the DSN arraying period. The time that they could reasonably devote to NASA's business was strictly limited.

Day by day, as the full disk of Neptune began to come within range of *Voyager 2*'s cameras, the images that emerged from the deep space connection between JPL and the spacecraft generated great excitement and astonishment among the science teams reviewing them at JPL. First came the apparition later called the Great Dark Spot (see figure 10.3), a huge stormy vortex that gradually began to appear on the face of Neptune. Next, high wispy clouds that streaked across Neptune's upper atmosphere made their appearance. Later several small moons appeared, and Neptune's largest satellite, Triton, began to fill the video screens at JPL. Its strange mottled surface led some to conjecture that it was a high cloud cover; others thought they might be seeing the actual surface of Triton. That question would be resolved later, as *Voyager 2* approached even closer to the satellite. One day, to everyone's surprise, the images showed a faint new ring that appeared to completely encircle the planet. "A ring around Neptune? Incredible!" said the scientists.

So Voyager's grand finale continued, as the new images kept coming from Neptune day after day without interruption. There were no failures on the spacecraft, and no outages in the Deep Space Network or its partners, in this memorable event

Late on August 24, *Voyager 2* crossed through the plane of the dark-colored, tenuous rings and swept over the surface of Neptune at a distance of 3,000 miles to disappear from Earth view behind the planet. As expected, the radio signals that, at light speed, were taking more than four hours to travel between the spacecraft's high-gain 4-meter antenna at Neptune and the DSN's 70-meter antennas on Earth, suddenly ceased. A few hours later the spacecraft emerged from the shadow of Neptune, reestablished radio contact with Earth, and began the final act of its grand finale, a close flyby of Neptune's largest satellite, Triton. Three days later it was over. Close-up images of Triton provided a fitting climax to the great performance. Chief scientist Edward Stone reflected the air of ex-

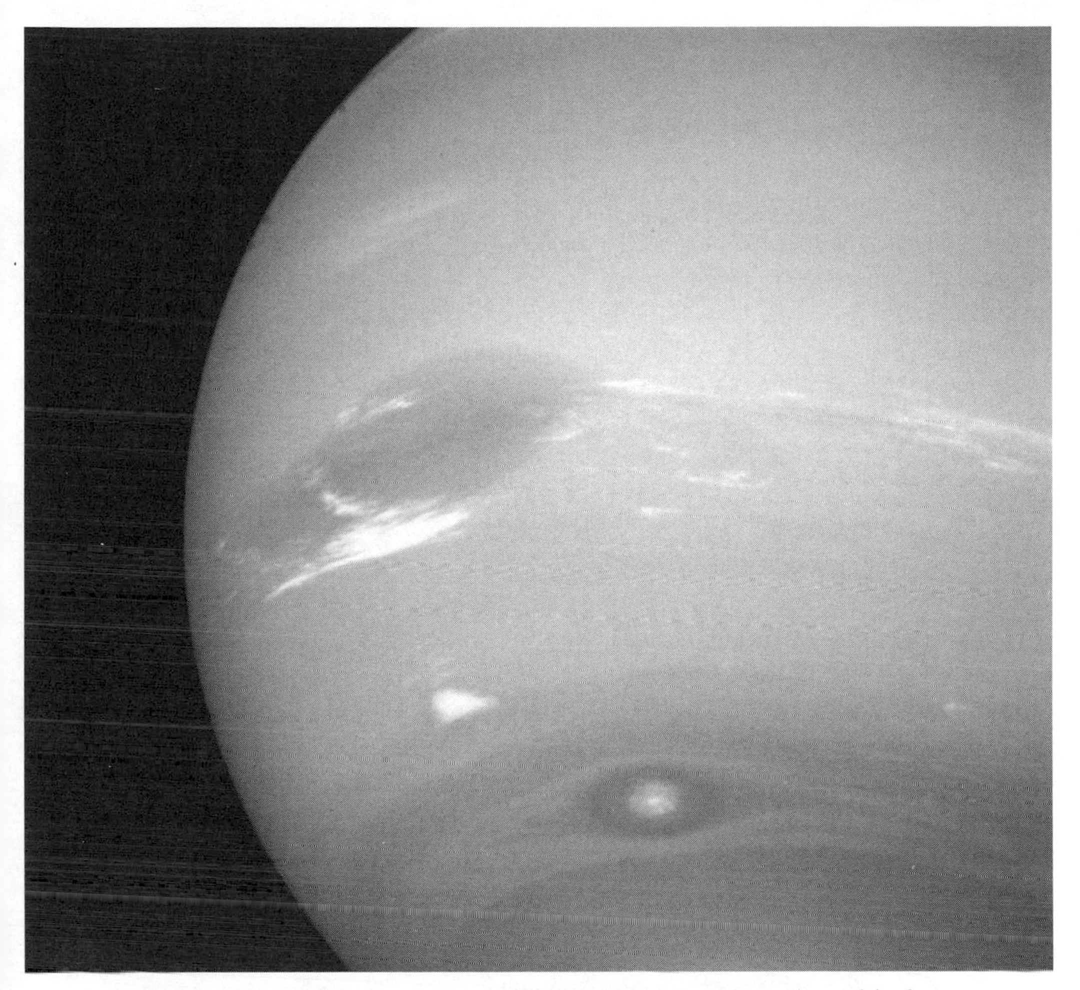

Figure 10.3. Prominent Surface Features of Neptune. This image shows three of the features that *Voyager 2* photographed as it neared the point of closest approach. The upper feature is the Great Dark Spot, accompanied by bright white clouds that undergo rapid changes in appearance. To the south of the Great Dark Spot is the bright feature that Voyager scientists nicknamed Scooter. Still further south is the feature called Dark Spot 2, which has a bright core. Each feature moves eastward at a different velocity, so they rarely appear close to each other as in this picture. NASA/JPL.

citement and astonishment that prevailed around JPL at the time. "The images [of Triton] returned this morning revealed a world unlike any we've ever seen," he said (see figure 10.4).

Press conferences at JPL brought the science and imaging results of *Voyager 2*'s encounter with Neptune to the public in (almost) real time and engendered a wealth of accolades for NASA from the world's media. Writing in *National Geographic* magazine in August 1990, Rick Gore titled his article "*Voyager*'s Last

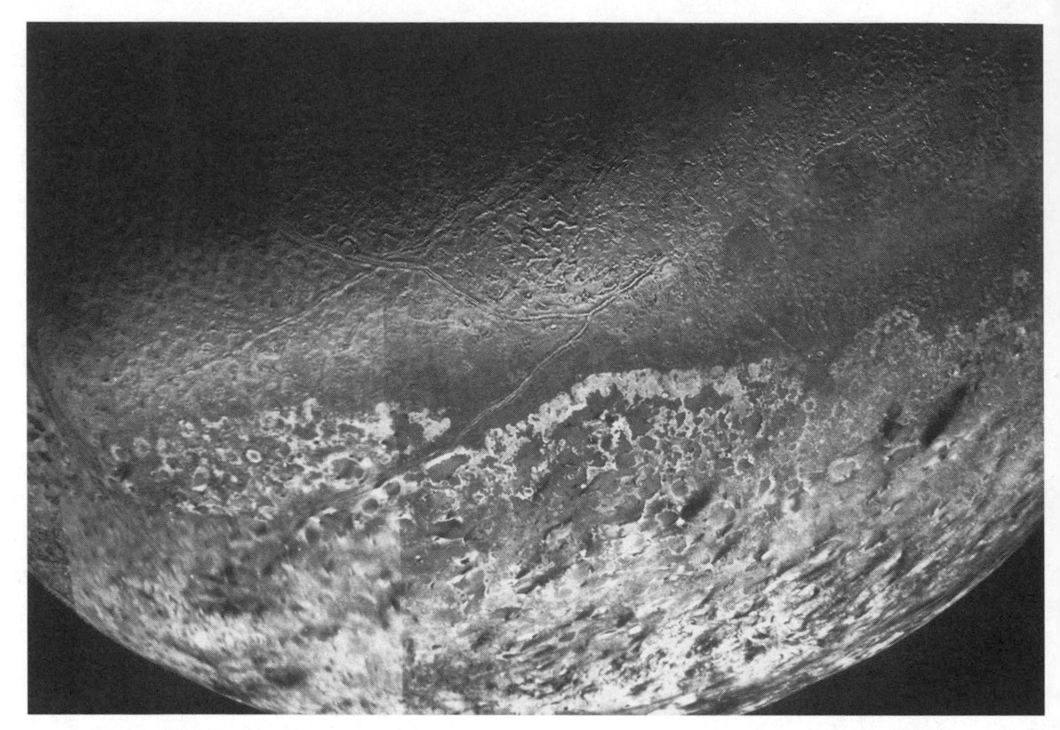

Figure 10.4. Surface of Triton. Taken as *Voyager 2* sped away from Neptune after its closest approach on August 24, 1989, this twenty-two-image mosaic of Triton shows fault valleys criss-crossing Triton's frozen surface near its equator. The valleys are filled by ridges of water ice extruded from the interior of the body. *Voyager 2* found Triton to be the coldest body in the solar system, with a temperature of –235°C (–391°F). NASA/JPL, USGS.

Picture Show." It was a most apt phrase to describe this event that he rightly called "both a first look and a last farewell."

Following its pass over Triton, *Voyager 2* looked back toward the Neptune system (figure 10.5) for a few last observations before heading off on an unending journey that would eventually take it clear out of the influence of Earth's solar system and into the void beyond. It had been, in the words of Brad Smith, leader of the Voyager Imaging Science Team, the "voyage of the century" (Smith 1990).

Both *Voyager 2* and its sister spacecraft *Voyager 1* had enough electrical power and thruster fuel to operate at least until the year 2020. By that time *Voyager 1* would be 12.4 billion miles from the Sun, and *Voyager 2* would be 10.5 billion miles distant—still within reach, albeit at considerably reduced communications ability, of the DSN's three Big Dishes.

Eventually both spacecraft will pass stars other than our Sun. Voyager navigators said that in 40,000 years *Voyager 1* would drift within 1.6 light years of AC+79 3888, a star in the constellation Camelopardalis. In some 296,000 years

Voyager 2 will pass Sirius, the brightest star in our sky, at a distance of about 4.3 light years (Massey 2003).

The Voyagers are destined to wander—perhaps eternally—through the Milky Way as a testament to the curiosity, ingenuity, and intelligence of the human beings that created them, and to the efficacy of the deep space connection that made the "voyage of the century" possible.

Bearing Down at Madrid, December 1989

On a Friday morning in December 1989, a shrilling telephone wakened Dale Wells from his predawn slumber. From years of exposure to similar disturbances, he instinctively knew that one of his antennas somewhere around the globe was in trouble. This time, though, he was not prepared for the news that waited for him an arm's length away. The caller was Alberto Manteca, the Spanish antenna maintenance engineer at the DSN Madrid complex, and the panic

Figure 10.5. Neptune with Triton. This beautiful image of Neptune, with the smaller crescent of its moon Triton in the background, was among *Voyager 2*'s parting shots as it began its infinite journey through the cosmic void. NASA/JPL.

in his voice conveyed the seriousness of the problem. It was midmorning in Madrid. About an hour earlier, said Manteca in rapid-fire Spanish, they had been making some elevation test runs for the recently launched *Galileo*'s mission to Jupiter when the new 70-meter antenna had suddenly come to a grinding halt. The antenna was stuck in the horizontal-pointing position, the right-side elevation bearing appeared to be out of alignment, and oil was pouring from the bearing housing. Any movement of the antenna, Manteca said, was accompanied by a "horrible cracking noise" (Wells 1999).

Neither engineer needed reminding of the critical role that the Madrid 70-meter antenna was to play in the Jupiter mission. After years of delay and redesign, the spacecraft *Galileo* had finally been launched into a convoluted trajectory to Jupiter by a solid rocket booster that was itself carried into Earth orbit by space shuttle *Atlantis* in mid-October. To reach Jupiter from this relatively low-energy launch, *Galileo* needed to use three complex interplanetary maneuvers known as gravitational assists, one from Venus and two from Earth. The date for *Galileo*'s first gravitational assist, the flyby of Venus, was set for February 10, 1990. The closest approach, some 10,000 miles above the surface of Venus, would occur at 5 hours 58 minutes 48 seconds Coordinated Universal Time.[1] The timing was precise, the maneuver was critical for the mission, and the Madrid 70-meter antenna was essential to its success. Now, barely eight weeks into its six-year mission, *Galileo* faced a situation that could have disastrous consequences, a situation that called for swift and decisive action (Mudgway 2001).

Two conflicting facts might have flashed through Wells's mind that dark morning. All of the elevation bearings on the three 64-meter antennas had worked perfectly from the time they were installed, Goldstone for about twenty-three years, Canberra and Madrid for about seventeen. On the other hand, one of the bearings at Madrid had failed only three years after that antenna had been modified to 70-meter diameter, though there had been no trouble with the other two antennas after their similar modification.

At this point, Wells knew, there was no time for further conjecture. The first order of business was to get the antenna up and running again as soon as possible, and that was clearly his responsibility. *Galileo* and other in-flight missions at JPL would have to work around the absence of one of the 70-meter antennas on which they depended for tracking and data acquisition support.

Speaking carefully in Spanish to avoid any misunderstanding, Wells instructed Manteca to slowly move the antenna up to its vertical position despite the hideous cracking and popping sounds that emanated from the bearing each time the antenna moved. Knowing that the bearing was destroying itself and probably doing irreparable damage, it was a risky decision, but typical of Wells's single-minded approach to solving problems for which he was responsible. He would justify his actions later, if anyone challenged them—but no one ever did. By the time the Madrid antenna was parked in the zenith position it was morning in California, and Wells called in to JPL to report the problem and the action he had already taken to get the antenna back into service. "I've got a spare bearing—it weighs almost five thousand pounds—and the heavy tools to handle it, at Goldstone. I'll get them to the airport today and on a plane to Spain, and then I'll go to Madrid to work the problem," he said. There were no objections. Two days later, when Wells arrived on-site at Madrid to begin the daunting repair task, the Spanish winter had set in, and working conditions high on the exposed alidade near the broken bearing were extremely uncomfortable and fraught with considerable danger.

Manteca had assembled a small team of Spanish mechanics to carry out the work under Wells's direction. They began by removing the end plate to expose the actual roller bearings. "When I saw the smashed bearings, my heart sank and I just wilted," Wells said. The dismal sight that confronted Wells is pictured in figure 10.6.

The immediate problem that confronted Wells was how to lift the weight off the elevation bearing so that he could extract the broken bearing and replace it with the spare he had shipped over from Goldstone. Determining the cause of the problem would come later. For now, he had to focus on getting the antenna back into service—*Galileo* could not wait for a review board to determine the proper course of action. Wells decided to fabricate a specially shaped steel saddle that would provide a flat working surface against which to jack up the elevation bearing, just enough to take the weight off the outer bearings. Relieved of the enormous weight of the tipping structure, the damaged outer bearing could easily, Wells hoped, be removed and replaced.

While Manteca found the necessary thick steel plate and cut it to shape, Wells set up the four hydraulic jacks that he had brought from Goldstone. With the shaped saddle in place and a total lifting capacity of 3,200 tons available, Wells began to increase the pressure in the hydraulic pumps that powered the jacks. It was a tense moment. Fred McLaughlin had joined Wells on-site by then and was watching the dials for any sign of movement that would indicate the broken bearing had lifted off its base. Wells recalled: "I was running the pumps, and when I got to 10,000 psi—that was the maximum rating for the pumps—there was still no call from Fred. I said maybe we need another jack but let's give it one last try, so I went up to 11,000 psi and suddenly it moved. We lifted it about one quarter inch and locked up the valves. That was the Friday afternoon [before Christmas] we all had the weekend off and spent Christmas in Madrid" (Wells 1999).

Figure 10.6. Cracked Elevation Bearing on Madrid 70-Meter Antenna, December 1989. With the antenna stowed in the vertical position for repair work, the exposed bearing reveals the extent of the damage. About half of the twenty-two rollers, as well as the inner and outer retaining rings that comprise the "race," are cracked. This bearing and its twin on the other side share the 4,000-ton load of the entire tipping structure. The central shaft is 24 inches (61 cm) in diameter, and each roller is about 3 inches (7.6 cm) in diameter. The hollowed core of the rollers is plainly visible. Photo by Dale Wells, NASA/JPL.

Years later, Fred McLaughlin remembered that Christmas in Spain, "sitting around in the Reyes Católicos Hotel in Madrid looking out at the rain on Christmas Day, and thinking that we were going to be here much longer than I expected. When it came to Christmas Day all the restaurants that we were familiar with were closed. We were all invited to the Officers Club at the U.S. Air Force base at Torrejón just outside of Madrid. We needed U.S. or Spanish passports to get in, but there was an Australian engineer from Canberra in the group, and his passport was not acceptable at the Officers Club—so none of us went. That group really hung together" (McLaughlin 1999).

Despite the promising start, the work suffered frustrating delays during the weeks that followed. Unusually bad weather stopped work completely on occasion, and several two-inch-wide bolts with threads frozen into their castings took days to extract before the damaged roller bearing races could be pulled off the huge shafts. Even that task required a custom-made hydraulic puller that had to be constructed on-site to handle the job. A special hydraulically powered machine was needed to press the huge new bearing back on the shaft, and it had to be airlifted to Madrid from the bearing manufacturer's plant in Sweden.

While these tasks had been simple enough when the gigantic elevation bearing assemblies were originally put together on the floor of the fabricator's plant in California, carrying out the same functions high on the alidade in a remote valley in Spain, with makeshift tools, while exposed to the vagaries of European midwinter weather, presented a completely different challenge. Everything about the task was unique, and procedures had to be created as the work progressed, for there was no precedent—nothing like this had ever happened in the network before.

Meanwhile, far across the solar system, *Galileo* moved ever closer to its appointment with Venus. That event would take place whether the DSN was ready or not, but utilizing it to redirect the spacecraft on its journey to Jupiter depended entirely on the perseverance, ingenuity, and dedication of this small group of workmen struggling to fix a broken antenna deep in the Spanish countryside.

Eventually the damaged bearing was pulled off the shaft and shipped back to JPL for analysis. The new bearing was pressed onto the shaft, aligned, and secured in place, and the jacks were slowly released to restore the antenna load to the new bearing. Test runs soon verified that the antenna was working correctly and, much to the relief of everyone associated with the mission to Jupiter, tracking and data acquisition support for *Galileo* resumed on January 26, 1990.

Two weeks later, on February 8, the DSN manager declared the Network was declared ready to support the *Galileo* encounter with Venus. No anomalies or exceptions were reported. Closest approach occurred at an altitude of 16,123 km at 05:58:48 UTC, February 10, 1990. The encounter period was covered both by DSS 63 and DSS 43, but bad weather in Australia degraded the quality of the telemetry data stream from that site. Telemetry from DSS 63 was excellent." *Galileo* received the gravitational kick that it expected, and the network resumed its hectic pace of planetary mission operations. The crisis was over (Mudgway 2001).

In the aftermath, JPL commissioned a special board to investigate the problem. The board found, "The most probable direct cause of failure was the fatigue breakdown of several rollers in the bearing." It recommended that all 70-meter antennas be retrofitted with roller bearings of a different type. This was done shortly afterwards under Wells's supervision, and there were no subsequent problems with the elevation bearings (Wood 1993).

Dale Wells received the Exceptional Achievement Medal from NASA (his second award) for his heroic efforts to correct the Madrid bearing failure. The citation read, "In recognition of the extraordinary field-engineering expertise in the hazardous work of replacing the elevation bearing of the Deep Space Network 70-meter antenna in time for the Galileo encounter with Venus."

Over the next couple of years, Wells gradually passed his responsibility and the benefit of his long experience to others at JPL. At a ceremony to mark his retirement in 1991, Robertson Stevens noted that Wells had been responsible for the maintenance and operational integrity of the Big Dishes since their construction in 1966, and observed that during the equivalent of more than sixty antenna-years of service, there were only two occasions on which any of the big antennas had to be taken out of service for more than a few hours for emergency repair due to design or construction problems. Those two occurrences involved the hydrostatic bearing problem at Goldstone in 1981 and the elevation bearing failure at Madrid in 1989.

It was a fitting tribute to a man whose career had, for twenty-five years, been devoted with consummate skill and passion to three of the world's biggest antennas, of whose history he had, by then, become an integral part.

Galileo and the Great Galactic Ghoul, April 1991

In the early days, the NASA space program occasionally experienced failures that defied all logical explanation. The cognoscenti in the labs and back offices of JPL took to attributing these inexplicable failures somewhat facetiously to the depredations of a Great Galactic Ghoul that lurked in deep space waiting to devour the flimsy creations sent by Earthlings to intrude upon its domain. With the passage of time, however, it seemed that the Great Galactic Ghoul had been subdued by the great advances in expertise, innovation, and technology with which the engineers from JPL endowed their later spacecraft, and awareness of its existence faded. The Ghoul, they thought, had gone away.

But the Great Galactic Ghoul had not gone away. It struck again in full force, against *Galileo,* in April 1991.

The setbacks with which *Galileo*'s mission to Jupiter had to contend in the decade or more prior to its launch were described, with considerable insight, by Bruce Murray, a former director of the Jet Propulsion Laboratory. In 1988 it appeared to him that "Galileo is hemmed in on every side by dark clouds, so towering that even the most innovative mission analysis and engineering may not be able to overcome them." Looking ahead, he said, "Perhaps Galileo will reach Jupiter in 1996 and start radioing back new secrets about the Sun's greatest companion. I hope so. But if it does not," he suggested, "perhaps that much modified Galileo spacecraft should finally be housed in the National Air and Space Museum. There it could serve as a billion-dollar monument to a time in America when political mediocrity triumphed over technical competence and dedication" (Murray 1989).

But the Galileo project, led by its tenacious project manager, John Casani, survived, and finally in October 1989 began what Bruce Murray called the "slow boat to Jupiter." It was scheduled to arrive there six years later. On arrival it would first release a probe to test Jupiter's atmosphere and then begin its own twenty-month-long orbital tour, returning imaging and other science data to Earth as it made one or more passes close by each of the Jovian satellites.

The vast amount of data accumulated by the orbiting spacecraft was to be transmitted to back to Earth at very high rates, up to 134 kilobits per second. Such high-rate telecommunications links from such a distance were made possible by the enhanced performance of the new 70-meter antennas of the DSN, and by operating the spacecraft radio transmitters at X-band through a furlable parabolic antenna 16 feet (4.8 meters) in diameter. This was known as the High Gain Antenna (HGA) to distinguish it from a second antenna carried on the spacecraft, the much simpler, fixed Low Gain Antenna (LGA), which operated only at S-band and was intended for use in the early part of the mission when communications distances were much smaller.

The HGA was an umbrella-shaped structure consisting of a fine mesh of gold-plated molybdenum wire held in shape by a set of eighteen flexible graphite-epoxy ribs. Through the early part of the mission, the HGA remained in its folded or stowed position, and the spacecraft communicated with the DSN over its LGA/S-band link. At the appropriate time the spacecraft was to be commanded to deploy the HGA, switch to X-band, and begin returning science data at the higher data rates over its HGA/X-band link. That was the plan as the long journey began at Cape Canaveral in October 1989.

The rationale for *Galileo*'s long and roundabout journey to Jupiter was discussed earlier. The convoluted shape of the path taken by the spacecraft is evident in figure 10.7.

Following the successful launch, a spirit of cautious optimism began to stir in the scientists, engineers, and controllers at NASA and JPL as they nursed the huge new planetary spacecraft through its carefully scripted sequences and each of the key events began to fall into place. The Venus encounter was executed

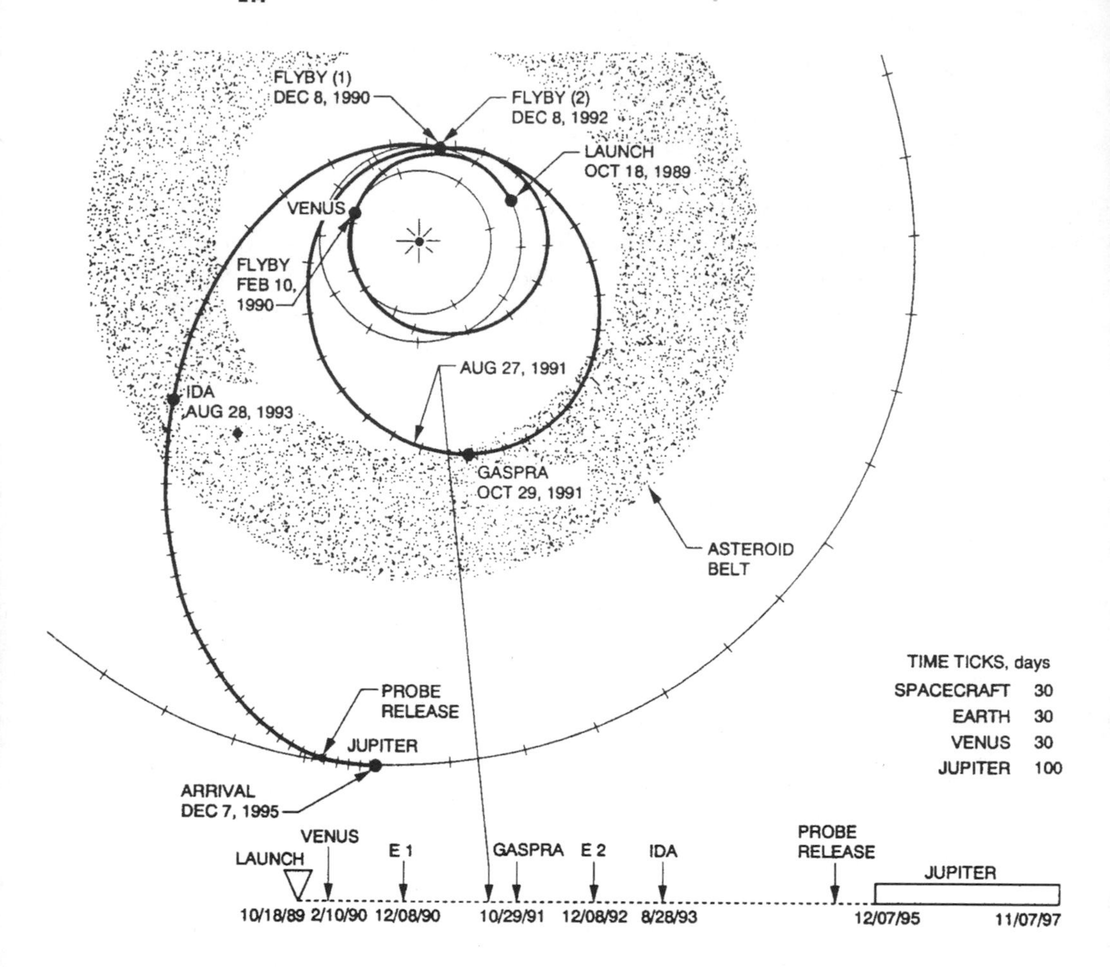

Figure 10.7 *Galileo*'s Venus-Earth-Earth Gravity Assist (VEEGA) Path. Key events in *Galileo*'s transit from Earth to Jupiter are identified on the lower timeline. They included close encounters with two asteroids, Ida and Gaspra. Following release of the probe into Jupiter's atmosphere, *Galileo* was inserted into an elliptical orbit around Jupiter from which it conducted scientific observations of Jupiter and its satellites. The primary mission was to end in November 1997. The original plan called for deployment of the HGA shortly after the first Earth encounter (E1). NASA/JPL.

perfectly and, imbued with its additional energy, the spacecraft headed back for its first close encounter with Earth. That too was executed flawlessly, and the navigators verified that the predicted boost of energy had been imparted to the spacecraft by Earth's gravitational forces. *Galileo* took spectacular pictures of the south polar region as it streaked diagonally across the South Atlantic Ocean and then, with all systems "go," began a another orbit of the Sun that would bring it back for a second Earth encounter two years later.

During this period when the spacecraft was close to Earth, the DSN tracked *Galileo* with its 34-meter antennas while the large antennas were released from service for short periods to replace the elevation bearings. By early 1991 both the spacecraft and the network were ready for the next phase of the mission, and engineers eagerly anticipated the deployment of the HGA and the switch-over to the better-performing X-band communications links.

Slowly and carefully the controllers worked through the sequences that preceded the "deploy HGA" command, and finally instructed the spacecraft to execute the command on April 11, 1991. It was expected to take about three minutes. The precise time at which controllers were to receive confirmation that the HGA was deployed came and went, but they saw nothing. The unthinkable had happened. The HGA had not deployed. The Ghoul had struck again.

Despite the gravity of the situation, the spacecraft engineering team remained calm. They quickly determined that three of the eighteen ribs had failed to release and were restrained in the stowed position. There was no structural damage, but the antenna was quite useless in a partially opened condition. While the mission continued in its former S-band/LGA arrangement, the project manager established an HGA Recovery Team to evaluate the problem and propose a course of action to free the stuck ribs. Various ideas of ever-increasing complexity were suggested and tried out over the ensuing months, to no avail. Experts were called in and erudite opinions were given, but at the end of the year the recalcitrant HGA remained stuck. Since the LGA/S-band system could not possibly provide useful data return to Earth from Jupiter distance, it began to look as though *Galileo* had been struck a mortal blow.

As had so often happened in such situations, engineers in the back offices and labs of JPL saw the loss of the high-performance HGA/X-band communications link on which original mission depended, not as a failure, but as a challenge to their ingenuity. They began brainstorming ways in which the *Galileo* mission might yet be saved. While attempts to free the HGA continued through 1992, DSN and spacecraft design engineers explored various options for further improving the S-band performance of the 70-meter antennas. By the time of the second Earth encounter in December, they had developed a plan that could indeed produce a viable mission. It would require in-flight changes to the spacecraft computer programs to improve the efficiency of the downlink telemetry transmissions, substantial enhancement of the DSN receiving capability at S-band, and a significant expansion of the data processing capability at JPL. It could be done, they said—but it would take time and money.

At that point there was little time and no money, but Bill O'Neill, the new project manager for *Galileo,* decided to carry the idea forward. Strongly motivated by his science team and supported by persuasive technical arguments from the communications experts at JPL, O'Neill went to Washington and succeeded in convincing NASA headquarters that he had a viable plan to recover the *Galileo* mission. The additional funds were made available, and O'Neill announced the new mission plan in February 1993.

"We are now proceeding to implement the Galileo Mission using the LGA. We are absolutely confident of achieving at least 70 percent of our primary objectives." He continued, "The success of Galileo without the HGA at Jupiter will be a technological triumph. Developing the extensive new spacecraft flight software and ground software and including enhancements in the DSN is a big challenge to complete in just three years. However, the Project and the DSN are fully up to the challenge. *Galileo* will indeed fulfill its promise and be a magnificent mission to Jupiter" (O'Neill 1993).

The new course was set and the new design work moved ahead. The technical details of the changes and additions that were required on the spacecraft and in the DSN were extremely complex and are fully discussed by O'Neill et al. (1993) and Mudgway (2001).

The demanding task of reprogramming the onboard computers while the spacecraft was in flight, hundreds of millions of miles away from Earth, was assigned to the spacecraft engineers and mission controllers. To take advantage of the greater efficiency in the S-band telemetry streams that resulted from reprogramming the spacecraft data system, the DSN would equip all three 70-meter antennas with new digital receivers and new data processing units that contained special decoding, recording, and signal-combining facilities. Because the Canberra site would have the best and longest view of Jupiter during *Galileo's* orbital tour, extra attention was paid to maximizing the Canberra 70-meter performance. The antenna was fitted with a special S-band ultra-low-noise receiving horn, and elaborate arrangements were made to combine the Canberra 70-meter antenna with all of that site's 34-meter antennas. To further boost the receiving capability at Canberra, the DSN called on its former partner in the *Voyager 2* Neptune encounter, the 64-meter antenna at Parkes, to join the array. And during overlap periods when Jupiter was in view of both Goldstone and Canberra, the 70-meter signal from Goldstone was also to be transmitted to Canberra and added to the array.

The end result of this fantastic concatenation of advanced data capture and manipulation technology was to increase the capacity of *Galileo*'s S-band data channel from Jupiter by a factor of ten, from about 10 bits per second to about 100 bits per second. At the time it did not seem like much, considering the effort it took and the fact that the original X-band data rate would have been more than a thousand times greater. However, that was the "bottom of the barrel"—there was no further capability left. Besides, there was another cause for con-

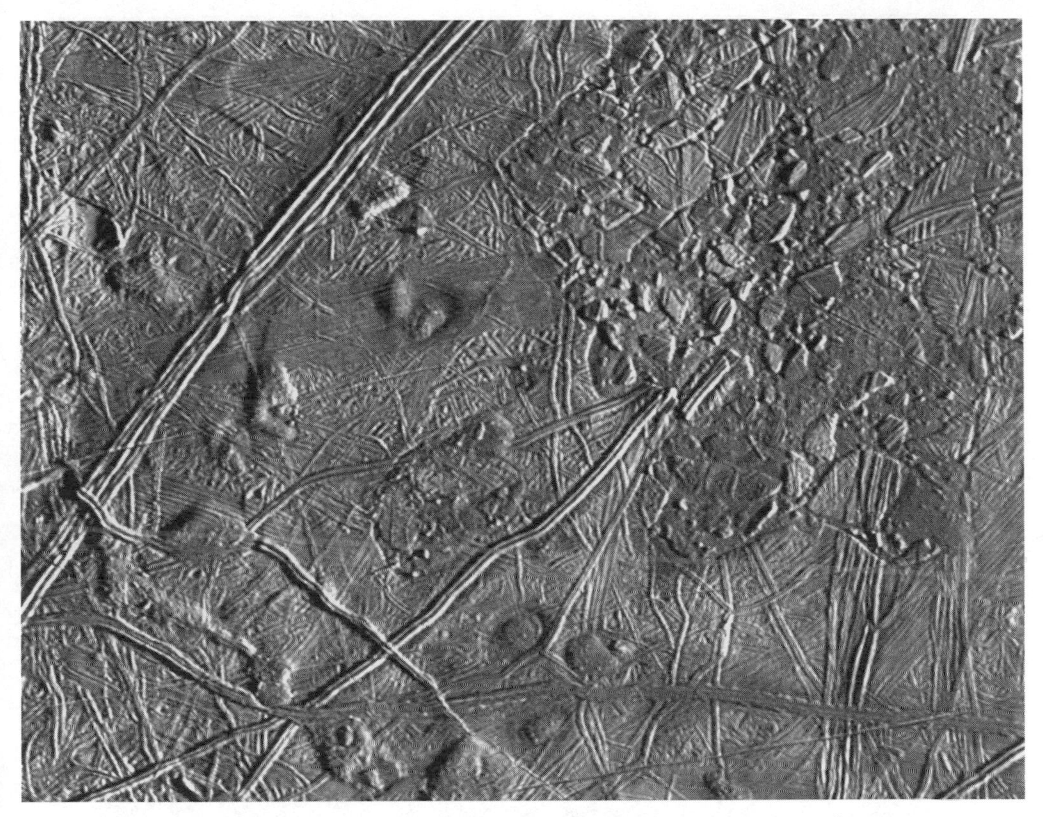

Figure 10.8. Icy Surface on Europa. This complex icy terrain on Jupiter's satellite Europa was photographed on *Galileo*'s sixth orbit around the planet in February 1997 from a distance of 17,900 kilometers (11,100 miles). The area shown in the picture is about 100 × 140 kilometers (62 × 87 miles) in extent. The upper right part of the picture shows terrain that has been disrupted by an unknown process superficially resembling blocks of sea ice during a springtime thaw. The smallest visible object is about 400 meters (0.25 mile) across. NASA/JPL, Arizona State University.

cern. On paper it seemed too complicated to work. Even if the engineers in Canberra succeeded once, could they continue to deliver the science data to JPL reliably, day after day, for the next two years?

As it turned out, the data rate was enough and the arrayed antennas did work as planned. Every day, the long-awaited imaging and science data from the wounded spacecraft trickled into JPL from Jupiter via the DSN/Parkes antenna arrays. Gradually the science data products accumulated into a massive repository of unique new information about the "court of King Jupiter," as William R. Newcott in the September 1999 *National Geographic* described the fascinating collection of satellites that orbit the giant planet.

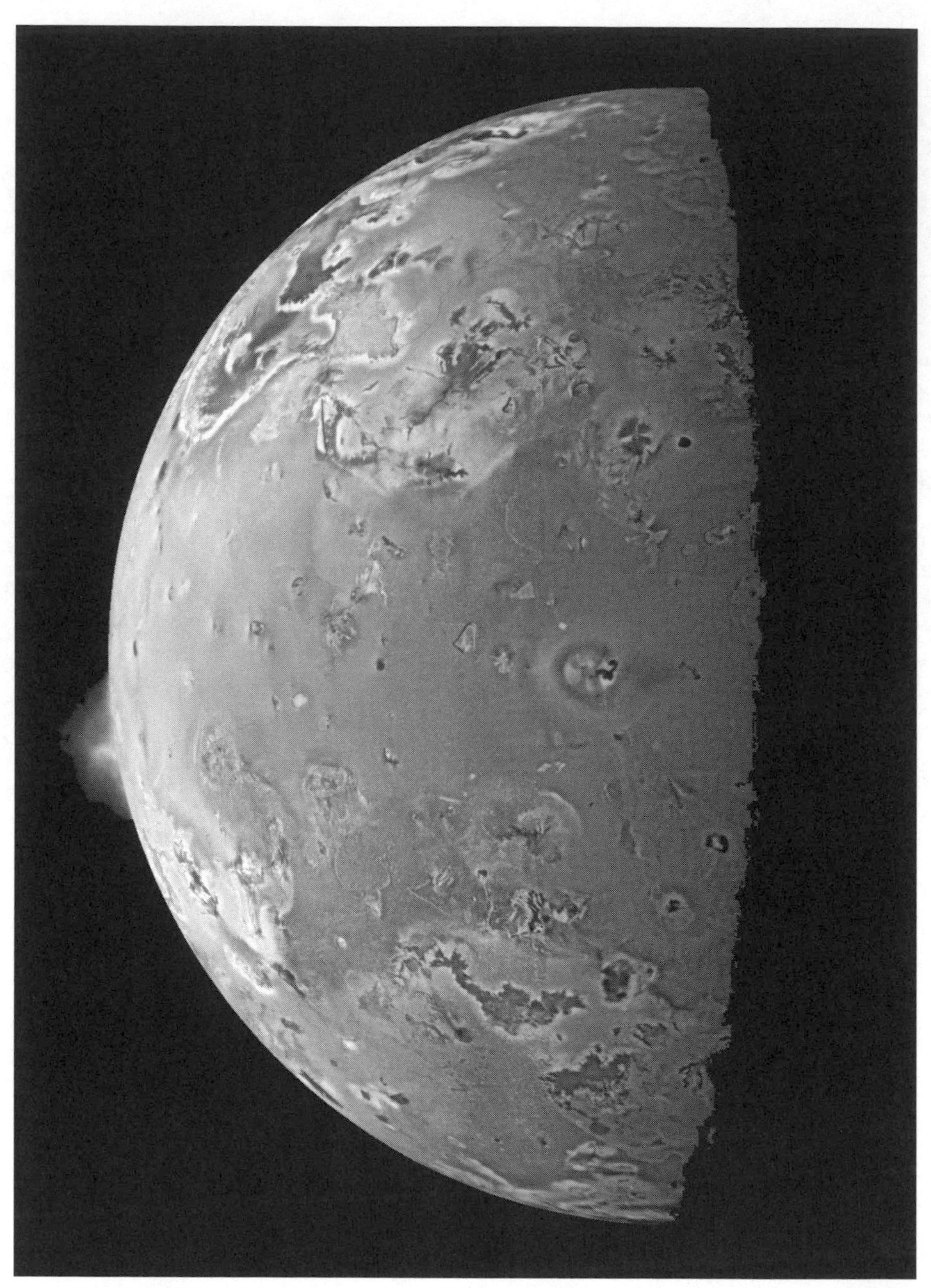

Figure 10.9. Volcanic Plumes on Io. This image of Io was acquired during *Galileo*'s ninth orbit around Jupiter, in November 1997. A volcanic plume on the bright edge of the satellite is erupting from a volcanic depression named Pillan Patera. The plume, 86 miles (140 kilometers) high, is bluish in color. *Galileo* observed several other volcanic plumes on Io as it viewed the satellite from different angles on later encounters. NASA/JPL, University of Arizona, Lunar and Planetary Laboratory.

Despite the low data rate of the S-band/LGA communications channel, the data products returned from the *Galileo* imaging system possessed astonishing clarity of detail, as the photographs of Europa and Io show in figures 10.8 and 10.9. These images were downlinked to Earth via the DSN/Parkes antenna array during multiple science-data playback sequences at 40–80 bits per second over five-week periods in 1997.

When the primary *Galileo* mission ended in December 1997, it was hailed as a great success throughout the scientific community. Much had been learned about Jupiter and its satellites in general and the unique nature and origin of volcanic activity on Io. The potential for discovery of water and possible extraterrestrial life forms beneath the icy surface of Europa became topics of particular interest. So great was the fascination with Europa that NASA decided to continue funding for an additional two years on what became known as the Galileo Europa Mission. The spacecraft continued to operate perfectly on the LGA and, within the network the technique of arraying multiple antennas became a routine matter. More science data was accrued as time went by. In fact, the Europa mission was so productive that, for little additional cost to NASA, it was extended a second time, as the Galileo Millennium Mission.

By the beginning of 2002, *Galileo*'s days were numbered. Because the supply of propellant needed to steer the spacecraft was almost exhausted, mission controllers at JPL feared that they might unexpectedly lose control of the spacecraft. To avoid the slightest chance of *Galileo* crashing into Europa and possibly contaminating it with Earthly organisms, NASA decided to send it to a predictable demise in the crushing pressure of Jupiter's dense atmosphere. The flight team aimed the spacecraft to skim just a hundred kilometers (62 miles) above Io's surface on January 17 on a ballistic trajectory that would enter Jupiter's atmosphere in September 2003. It was hoped that *Galileo*'s imaging system would survive yet another dose of radiation from the planet as it made its last pass over Io, and would return what could be called *Galileo*'s Last Gasp. And *Galileo* did survive, faithfully returning data right to the end.[2] JPL announced the spacecraft's demise in a news release on September 21, 2003. "The Galileo spacecraft's 14-year odyssey came to an end on Sunday September 21 when the spacecraft passed into Jupiter's shadow and then disintegrated in the planet's dense atmosphere at 11:57 a.m. Pacific Daylight Time. 'We haven't lost a spacecraft; we've gained a stepping stone into the future of space exploration,' said Dr. Torrance Johnson, Galileo project scientist."

And what of the Great Galactic Ghoul? Perhaps it had met its match after *Galileo* survived its near death blow in 1991, and went looking elsewhere for victims. Perhaps it lurked near Mars, where *Mars Observer* perished in 1993 and both *Mars Climate Orbiter* and *Mars Polar Lander* were lost in 1998. Again perhaps, as William Burrows (1998) suggested, it moved away to wreak havoc on the Russian space program. Who knows?

Midgets on Mars, July 1997

Toward the end of the first President Bush's administration, a weakening of congressional support for "big science" items in NASA's budget had begun to threaten plans for future planetary exploration. Most such plans required ever-larger spacecraft and launch vehicles for more ambitious and therefore more costly missions. When Dan Goldin became NASA administrator in 1992, he quickly recognized the compelling need to reduce the cost of planetary missions if NASA's program of solar system exploration was to survive. The time had passed for hugely expensive missions such as *Galileo, Viking,* and the yet-to-be-launched *Cassini* mission to Saturn. What was needed now, suggested Goldin, was a plan for planetary missions with a maximum development cycle of three years and a maximum development cost of $150 million. With that concept as its basis, Goldin called for a new approach to the design of planetary missions. It would be called the Discovery Program and would be characterized by the catchphrase "Faster, better, cheaper." Goldin set out to promote the "Faster, better, cheaper" concept throughout NASA and challenged the science community to propose new initiatives for the Discovery Program (McCurdy 2001).

JPL responded with a proposal for a Mars exploration program that used a small, lightweight spacecraft capable of performing a high-interest Mars mission with sharply focused but clearly limited objectives at low cost.

Among a great many innovative approaches to reducing the size and weight of the spacecraft, designers eliminated the large, heavy antennas and transmitters, since the advanced capabilities of the DSN's 70-meter antennas would compensate for the lesser capabilities of smaller and lighter antennas aboard the new spacecraft. The resultant saving in spacecraft weight translated directly into Goldin's three-word mantra. The midget spacecraft not only was faster, better, and cheaper than its Viking predecessors but would also support high-performance uplink and downlink communications at Mars distance by using the 70-meter antennas of the DSN. It was to be called *Mars Pathfinder.*

On December 4, 1996, a Delta launch vehicle blasted *Mars Pathfinder* into a fast-track trajectory to Mars with a planned arrival date of July 4, 1997. *Pathfinder* landed on the forbidding surface of Mars precisely on time and on target. Retarded by parachute and cushioned by inflatable airbags, the spacecraft survived the descent and landing impact without damage and, a few hours later, began transmitting data to JPL from the Martian surface at 6 kilobits per second. The event marked NASA's return to Mars after more than twenty years.

The *Pathfinder* spacecraft then split into two space vehicles, a lander and a rover. The lander not only carried the rover—a robot the size of a microwave oven—but also contained a package of scientific instruments to investigate the atmospheric structure and meteorology of Mars, a spectrometer to determine the composition of the rocks and soil, and a versatile color television camera. The rover was likewise equipped with a spectrometer and a color camera and was designed to explore the Martian surface in the near proximity of the lander. A UHF radio link transferred commands and data back and forth between the battery-powered rover and the solar-powered lander. X-band uplinks and downlinks between the lander and the DSN allowed controllers at JPL in Pasadena to navigate the rover on the Martian surface in real time, and to command the lander to perform various observations, including imaging, and return the data so gathered to scientists at JPL.

Images of the rover taken by the lander and vice versa, plus engineering data from both, confirmed that the two vehicles were in their proper positions and in good operating condition. With the downlink telemetry data rate set to 6300 bits per second, science activities from the surface of Mars began in earnest. The data delivered by the DSN stations to JPL in real time were subsequently processed into images by the Mars Pathfinder Science Team and made available to the world media, and to the general public on the World Wide Web.

Downlink sessions with the lander were of short duration, generally about ninety minutes, during which time about 60 megabits of science and engineering data were returned when communication links allowed.[3] By the time *Pathfinder* had completed its primary thirty-day mission on August 3, it had returned 1.2 gigabits of data including 9,669 images of the Martian landscape.

The images returned from the *Pathfinder* lander and rover were remarkable indeed, but it was the midget, semiautonomous rover that captured people's imagination. To accommodate the great public interest in following the mission via the World Wide Web, JPL engineers in cooperation with several educational and commercial institutions constructed twenty *Pathfinder* mirror sites. Together these sites recorded 565,902,373 hits worldwide during the period July 1–August 4. The highest number of hits occurred on July 8, when a record 47 million hits were logged, more than twice the number received by the official Web site for the 1996 Olympic Games in Atlanta.

A typical high-quality *Pathfinder* image, transmitted from Mars to Earth over the DSN telemetry downlink, is shown in figure 10.10. The inherent capability of the DSN downlink contributed to the remarkable detail observed in this and other *Pathfinder* images.

At an October 8, 1997, press briefing at JPL, *Mars Pathfinder* project scientist Dr. Matthew Golombek described what facts the scientists were seeing in the data coming back from Mars, and what conclusions might be drawn. The evi-

Figure 10.10. *Mars Pathfinder* View of Mars Surface, July 1997. This picture, taken by the Imager for *Mars Pathfinder* (IMP) camera on July 18, 1997, shows the remotely controlled rover *Sojourner* as it placed its Alpha Proton X-ray Spectrometer upon the surface of the Martian rock named Yogi. Early analysis of the Yogi data showed it to be low in quartz content, somewhat like the common basalts found on Earth. NASA/JPL, University of Arizona, Lunar and Planetary Laboratory.

dence seemed to reinforce theories that water once existed in stable form on Mars. As water is the principal ingredient required to support life, Golombek said, that "leads to the sixty-four-thousand-dollar question: Are we alone in the universe? Did life ever develop on Mars? If so, what happened to it, and if not, why not?" The following day an official press release reported that "Mars was appearing more like a planet that was very Earth-like in its infancy, with weathering processes and flowing water that created a variety of rock types, and a warmer atmosphere that generated clouds, winds, and seasonal cycles" (JPL 1997).

Communications with the lander continued with no sign of trouble until September 27, when the Goldstone 70-meter station was unable to detect a downlink at the scheduled transmission time. There was conjecture that, without the heat generated by the battery-powered transmissions, the spacecraft temperature had fallen below its operating limit and its computers would no longer function correctly. Whatever the cause, the downlink was never recovered, and further efforts were discontinued in mid-October.

At the time of the last *Pathfinder* downlink, the lander had operated for nearly three times its design lifetime of thirty days, and the rover had operated for twelve times its design lifetime of seven days. The 2.6 billion bits of science data that had been returned from Mars since the landing on July 4 included more than 16,000 images from the lander and 550 images from the rover, as well as more than fifteen chemical analyses of rocks and extensive data on winds and other weather factors.

The cost of designing and building the lander had been $171 million, and for the rover $25 million. Their combined costs would have been a mere round-off error in the $3 billion cost (in 1997 dollars) of NASA's previous mission to Mars, the Viking landers in 1976 (McCurdy 2001). The designers of the spacecraft and the mission had every reason to be proud. "This mission demonstrated a reliable and low-cost system for placing science payloads on the surface of Mars," said Brian Muirhead, *Mars Pathfinder* project manager at JPL. "We've validated NASA's commitment to low-cost planetary exploration, shown the usefulness of sending micro rovers to explore Mars, and obtained significant science data to help understand the structure and meteorology of the Martian atmosphere and to understand the composition of the Martian rocks and soil" (Mudgway 2001).

NASA also seemed well satisfied. In a formal press release issued a few weeks later, Administrator Daniel S. Goldin expressed NASA's appreciation for the efforts of all involved with the *Mars Pathfinder* mission. He said, "I want to thank the many talented men and women at NASA for making the mission such a phenomenal success. It embodies the spirit of NASA, and serves as a model for future missions that are faster, better and cheaper. Today, NASA's Pathfinder team should take a bow, because America is giving them a standing ovation for a stellar performance" (Mudgway 2001).

Although it was not mentioned explicitly, those sentiments implicitly included praise for the Deep Space Network and its deep space connection.

Toward the Future

In 1998 a group of eminent telecommunications engineers at JPL were assigned a task that was directed specifically toward the future. The goal was to identify the most cost-effective options for the growth of deep space telecommunications. They began by rationalizing a basis for the study:

> With the advent of faster, cheaper planetary missions, the coming decade promises a significant growth in the number of missions that will simul-

> taneously be supported by NASA's Deep Space Network. In addition, new types of missions will stretch our deep-space communication capabilities. Ambitious outer planet missions, with extremely tenuous communications links due to their great distances, and data-intensive orbiter or in situ missions incorporating high-bandwidth science instruments, will demand improved telecommunications capabilities. Ultimately, our ability to create a virtual presence throughout the solar system will be directly linked to our overall deep-space telecommunications capability. (Edwards et al. 1999)

It was a neat, concise statement of the situation that faced the network as it entered the new millennium. The team's findings included a strong emphasis on the benefits that would accrue from the addition of Ka-band (32 GHz) communications to the network's current communications channels on S-band (2.3 GHz) and X-band (8.4 GHz). "A Ka-band mission," they asserted, "can return four times more data than a comparable X-band mission." Furthermore, the four-times improvement factor could be traded off in various ways between the spacecraft or the ground station antennas, or it could be used to reduce the tracking time required to deliver the total data package, or some combination of all three options. It was an attractive proposition. Their recommendations conveyed a sense of immediacy. "Add Ka-band communications capability to five of the 34-meter antennas by 2003, three more by 2006, and, finally, implement Ka-band on the three 70-meter antennas by 2009," they suggested (Edwards et al. 1999).

These ideas are illustrated in table 10.1. The last row of the table shows the telemetry data rate that could be supported by the DSN's 70-meter and 34-meter high-efficiency antennas from JPL's "baseline" spacecraft at Jupiter distance using X-band and Ka-band downlinks. The data show that the Ka-band link supports telemetry data rates that are at least three to four times higher than those supported by the existing X-band links.

A veteran DSN telecommunications expert, Charles T. Stelzried, explained the ideas embodied in the table:

> These data assume the antenna-specific parameters given in the first column. Based on long experience with planetary telecommunications systems, they represent realizable values of theoretical estimates of these parameters when unavoidable losses due to ground antenna pointing error and atmosphere-related factors are taken into account. Additional losses accrue from spacecraft pointing error, parameter error margins, acceptable bit error rates, coding, modulation, antenna elevation, etc. These factors combine to reduce the realizable data rate advantage of Ka-band to

less than the sixteen-fold value that is represented by the four-times increase in Ka-band frequency relative to X-band.

For the purposes of this study, JPL defined a "baseline" spacecraft to have an antenna of 1.4 meters diameter with an efficiency of 50 percent, together with X-band and Ka-band transmitters each with an RF power output of 10 watts. This was considered to be representative of planetary spacecraft of the future, where the constraints of mass, power, volume and cost would be more stringent than in the past. In its link budget design, this study also embodied certain other assumptions about telemetry data coding, weather losses, antenna elevation angle, bit error rate, etc., all of which remained constant for each of the examples that were evaluated for comparison of their relative performances. (Stelzried 2003)

The data rates given for the baseline spacecraft at Jupiter can easily be extrapolated to other scenarios by applying an appropriate multiplying factor. For example, the data rates for the baseline spacecraft at Mars, Saturn, and Uranus—distances of about 1.5, 10, and 20 AU—are obtained by multiplying the Jupiter data rates by factors of about 10, $^1/_4$, and $^1/_{16}$ respectively. These factors correspond to the inverse square of the ratio of the other planet's distance to Jupiter's distance (approximately 5 AU) from Earth. Similar multiplying factors for spacecraft antenna size and efficiency, or spacecraft transmitter power, could be derived to effect trade-offs in parameters other than range.

This study, however, focused on the issue of Ka-band versus X-band. When converted to data volume returned,[4] the enhanced Ka-band data rates were used to effect trade-offs between the overall cost of upgrading the existing antennas to Ka-band and the cost of implementing additional 34-meter X-band antennas, in order to support an ever-increasing ensemble of planetary spacecraft.

At the time, engineers foresaw no significant technical obstacles to imple-

Table 10.1. Projected Downlink Data Rates on DSN Antennas for Spacecraft at Jupiter Distance

Antenna type	70-m diameter		34-m diameter	
Frequency	X-band	Ka-band	X-band	Ka-band
Efficiency*	0.67	0.39	0.74	0.56
System noise temperature (T), K	22.2	52.3	24.7	54.3
Antenna gain (G), dB	74.1	83.3	68.2	78.6
Figure of merit (G/T), dB	60.6	66.1	54.3	61.3
Data rate from Jupiter, kbps	80.1	253.5	18.7	83.4

*Theoretical (in vacuum) value adjusted for atmosphere and ground antenna pointing losses.

menting these recommendations on the 34-meter networks, since Ka-band technology had been under development by researchers at JPL for some time, but implementing Ka-band on the 70-meter antennas was perceived as a difficult, and possibly insurmountable, technical challenge.

The problem was a familiar one: the reflecting surface "drooped" when the antenna moved between its vertical and horizontal positions. Engineers called this defocusing effect "gravitational deformation." They had significantly reduced its performance-degrading influence on S-band and X-band operation by stiffening the reflector backup structure, as we saw earlier. Operating the Big Dishes at 32 GHz, however, was an entirely different matter. As engineers perceived the problem, "Making the 70-m antennas perform well at the one-centimeter wavelength of Ka-band will be technically challenging, as the primary antenna surfaces deform by a significant fraction of this wavelength due to changing gravitational loads as the antennas track targets in elevation angle" (Edwards et al. 1999).

As intractable as the problem appeared to be in 1999, researchers were already working on two approaches to compensate for the loss of performance caused by gravitational deformation. One approach added a programmed "deformable mirror" to the microwave optics arrangement to compensate for the wavefront distortion, while the other relied on an electronic manipulation of the signals collected by a cluster of microwave feeds. Where those efforts would lead remained to be seen, but it was clear that DSN planners intended the 70-meter antennas to be part of the network's strategy well into the future.

That line of reasoning, however, raised difficult questions. How long would the existing 70-meter antennas last? What could eventually replace them? At the time of the study, the Big Dishes had been in continuous operation, in one form or other, for nearly thirty-five years. During that time they had accumulated an enviable record of achievements, one that was evident for the whole world to see. But by the start of the new century, the cost of operating and maintaining them at the level of performance required for deep space missions had risen, replacement parts were difficult and sometimes impossible to obtain, and new technologies such as multiple-antenna arrays, Ka-band communications, beam waveguide techniques, and adaptively controlled reflector surfaces offered significant advantages for large microwave antenna applications. For all these reasons, it was an appropriate time for the DSN to consider a long-term replacement for its 70-meter antennas.

Forty years earlier, JPL engineers had faced a similar situation with the forerunner of the modern DSN. Then, the new 64-meter radio astronomy antenna at Parkes, Australia, had afforded a useful source of technical ideas and hands-on

Figure 10.11. Green Bank Radio Telescope. The 100-meter GBT at NRAO, Green Bank, West Virginia, is the largest and most complex single-dish radio telescope ever built. A wheel-and-track design allows the telescope to view the entire sky above 5 degrees elevation. The main reflector surface, almost 8,000 square meters (85,000 square feet) in area, consists of 2,004 panels individually controlled to maintain the correct surface shape at all elevation angles. The high degree of surface accuracy thus obtained enables operation over a wide range of radio wavelengths, 3 meters through 3 millimeters (approximately 100 MHz through 100 GHz). An off-axis feed arm obviates the need for support structures that would otherwise degrade antenna performance. The weight of the moving portion of the antenna is 7,300 metric tons (16 million pounds). Photo by Mike Bailey, courtesy NRAO/AUI/NSF.

experience. Would the DSN turn once again to the radio astronomers as a point of departure in its search for a viable replacement for its 70-meter antennas, and what lessons might be learned from their recent experience? If they took that approach, the huge new radio telescope at the National Radio Astronomy Observatory at Green Bank, West Virginia, would be a good place to start.

The NRAO is a facility of the National Science Foundation and was operated under cooperative agreement by Associated Universities Incorporated. Construction of the Green Bank Telescope (GBT) was completed in August 2000, when it was dedicated as the Robert C. Byrd Green Bank Telescope. The immense structure, pictured in figure 10.11, claimed title to being the world's largest fully steerable radio telescope.

It was no coincidence that the project manager for the antenna construction portion of the huge new facility was the same Robert D. Hall who, as an engineering manager for the Rohr Corporation, had been responsible for construction of the giant 64-meter Advanced Antenna at Goldstone in 1963.

The GBT 100-meter antenna and the DSN 70-meter antenna were designed for entirely different purposes. One was better at mapping cosmic radio sources, the other at tracking planetary spacecraft; one was designed to receive only, the other to transmit and receive. The GBT used an active-surface, laser-based metrology system to maintain the antenna efficiency at all elevation angles and temperatures, an offset feed to avoid undesirable aperture blocking effects, multiple receivers for rapid switching between operating wavelengths, and a wheel-and-track system for the azimuth axis mounting. All were items of significant interest in the context of DSN experience with the 70-meter antennas.

To what extent and in what form any of these items would eventually find their way into future DSN designs was, of course, entirely for the DSN designers to decide. But if past experience was any indicator, they would assuredly look back for inspiration before moving forward to a design for the future.

In the meantime, there was little doubt that, whatever form the communications systems of the future might take, the DSN's "deep space connection" would continue to be an essential element of humankind's unrelenting quest to more fully comprehend the solar system and, ultimately, to discover the origins of life itself.

Appendix 1

An abridged summary of the critical performance specifications for the Advanced Antenna System, as presented to the American Association of Mechanical Engineers in November 1961, is given here.

Critical Performance Specifications for the NASA/JPL Advanced Antenna System, 1961

Item	Specification
Effective aperture area at 2.1 to 2.3 GHz frequency	Approximately ten times greater than existing efficient 85-foot antennas
System noise temperature at 2.1 GHz to 2.3 GHz frequency	10–15 kelvin for the antenna plus zenith atmospheric temperature; 3 kelvin for microwave feed and antenna reflector losses
Absolute pointing accuracy	±0.02°
Relative tracking accuracy with efficient, low-noise tracking feed	±0.02°
Angular coverage	Approximately half hemisphere toward equator within 5° of horizon
Angular rates and accelerations for pointing and tracking	Adequate to track twice-sidereal-rate targets with 3° cone of silence at zenith; able to provide tracking and pointing accuracy down to zero rate
Slew/scan rates and accelerations	0.25°/sec, 0.2°/sec^2
Wind (average over antenna surface)	0–30 mph: full pointing accuracy 30–45 mph: accuracy degraded by factor of 2 45–60 mph: accuracy degraded by factor of 4 Up to 70 mph: antenna drive to stow position along any path 70–120 mph: antenna stowed Up to 120 mph: no damage to antenna
Temperature	0°F to 135°F (–18°C to 57°C)
Sun	Partially or completely exposed
Ice, snow, with antenna in stowed position (survival)	One foot of snow, or one inch of ice with 60 mph wind

Appendix 2: Merrick's Hard Core Team

The impressive diversity of specialized technology that was represented by Merrick's Hard Core Team is evident in the following list of engineers and their areas of cognizance.

Pedestal, instrument tower, alidade, site facilities
  Ronald Casperson
Azimuth bearings
  Horace P. Phillips
Elevation bearings, azimuth and elevation drives, and gears
  Fred D. McLaughlin
Reflectors and backup structure
  M. Smoot Katow
Quadripod and feed cone structure
  Kenneth P. Bartos
Servo and master equatorial subsystems
  Robert J. Wallace
Analytical modeling for servo and master equatorial subsystems
  Houston D. McGinness
Optical and measurement techniques
  Chris C. Valencia
Resident engineer for industrial contractor
  David Ireland
Resident engineer at Goldstone site
  Donald H. McClure
Microwave design and analysis
  Philip D. Potter
Project administrator
  Paul C. Doster Jr.

JPL also appointed a contract administrator, Wallace P. Lord, and a quality assurance manager, Joseph P. Frey, to support the team's technical expertise in matters of procurement and quality control associated with the project.

Notes

Chapter 1. Introduction

1. Originally the collection of stations was called the Deep Space Instrumentation Facility (DSIF). Later it became the Deep Space Network (DSN).

2. A full account of the use of the Deep Space Network as a scientific instrument is given in *Uplink-Downlink* (Mudgway 2001).

Chapter 4. Reaching Further Out

1. The Deep Space Instrumentation Facility (DSIF) was renamed the Deep Space Network (DSN) in December 1963.

2. The DSIF required three stations to maintain continuous contact with the space probes as the Earth rotated, as explained earlier.

3. When the antennas were first built, English units were used to describe their diameters. By 1960 the DSIF had adopted the metric system and the 85-foot antennas became 26-meter antennas.

4. By the time the antenna was built, the location of the third antenna had been moved to Spain, but the environmental specification was essentially unchanged.

5. An abridged version of the critical performance specifications for the Advanced Antenna System (210-foot antenna) is shown in appendix 1.

Chapter 5. The Contract

1. As mentioned, the name Deep Space Instrumentation Facility (DSIF) was changed to Deep Space Network (DSN) in December 1963.

2. A list of the engineers that comprised Merrick's Hard Core Team is given in appendix 2.

3. The formal document that defined a basic conceptual design for the new antenna was Engineering Planning Document 5 (Merrick, Stevens, and Rechtin 1963).

4. The diameter was limited to 73 meters (240 feet), and the ratio of focal length to diameter was limited to the range 0.25 to 0.42.

5. The alidade structure is the rotating structural framework on which the reflecting part of the overall antenna is mounted. The alidade rotates in azimuth; the reflector rotates in elevation.

6. "Reducer" is an engineering term for a gearbox that translates high-speed, low-torque motion at an input drive shaft to low-speed, high-torque motion at an output drive shaft.

7. Although the name change to DSN was not officially announced until December 1963, it had come into common usage by this time. It will be used henceforth in this book.

Chapter 6. Goldstone, California

1. The Slab was a special assembly area set up near the Rohr plant in San Diego specifically for the purpose of fabricating, assembling, and fitting the tipping parts of the antenna before they were trucked to Goldstone.

2. The Cassegrain design is derived from an optical system used in most large telescopes to bring the focal point of the parabola close to the physical center of the parabolic surface. This allows a camera or, in the case of a radio antenna, a receiver to be placed at a point that is easily accessible.

3. The term "ultrasensitive receiver" is used here to describe what is really the low-noise maser front end of the receiver. Masers are described later.

4. Stretch-forming was a special technique common to the aircraft industry, where a sheet of aluminum was stretched over an appropriately shaped mandrel to permanent deform the metal sheet to the contour of the mandrel. This process, at which Rohr excelled, enabled sheet metal to be formed into complex three-dimensional shapes with great accuracy.

5. Because its position-sensing element was attached to the back of the main reflector surface, the master equatorial system provided a far more accurate estimate of the true position of the radio beam than the more common alternative of angle readout encoders attached to the shafts of the elevation and azimuth bearings.

6. One arcsecond is an angular measure approximately equal to 0.0003 degrees; 1,296,000 arcseconds equal 360 degrees of arc.

7. S-band refers to the radio frequency band around 2.3 GHz having a wavelength of about 13 centimeters. X-band refers to the radio frequency band around 8.4 GHz having a wavelength of about 3.6 centimeters.

8. Communication engineers use a logarithmic scale to measure the comparative performance of electronic devices. The units of this scale are called *decibels* (dB). In this scale, 3dB represents a factor of two, 6 dB a factor of four, and 9dB a factor of eight. 8 dB corresponds approximately to a factor of six and a half.

9. Orbiting the Sun, *Pioneer 6* was the first of two Pioneer spacecraft to investigate the interplanetary environment between Earth and the Sun. Later Pioneers investigated the interplanetary environment outward from Earth's orbit.

10. "Solar occultation" is an astronomical term used to describe the situation wherein an interplanetary spacecraft passes behind (is obscured by) the solar disk as viewed by a radio antenna on Earth. The spacecraft and its radio signal are said to be "occulted" by the Sun.

Chapter 7. Soon There Were Three

1. Bathker's unreserved praise for the remarkable performance of the 64-meter antenna was as much for its agility and beam-pointing control as for its radio-frequency capabilities. The former qualities were, in no small measure, a consequence of Merrick's prescient decision to incorporate the ME system of antenna-pointing control into the original 64-meter design requirements.

2. About this time the DSN adopted the metric system to describe its antennas. The 210-foot antenna became the 64-meter antenna; the 85-foot antennas became 26-meter antennas. Henceforth, the metric system will be used in discussing them.

3. The now familiar initials DSN for Deep Space Network, having replaced the term DSIF in December 1963, will be used hereafter.

4. NASA closed its deep space tracking station code-named DSS-51 near Johannesburg, South Africa, in June 1974. However, the antenna and much of the electronics remained in place and were subsequently used very productively for radio astronomy research by scientists of the South African National Institute for Telecommunications Research.

5. The DSN used a numerical sequence based on geographical longitude to identify its stations. Stations at the Goldstone longitude were designated DSS-11, DSS-12, and DSS-13. At the Australian longitude the stations were DSS-41 and DSS 42, in South Africa DSS-51, and in Spain DSS-61 and DSS-62. Under this system the new 64-meter stations became DSS-14, DSS-43, and DSS-63 respectively. From here on these identifiers will appear frequently in the narrative.

6. A detailed account of the influence of this and the other eras of planetary exploration on the operation of the Deep Space Network is given in Mudgway (2001).

Chapter 8. Bigger Is Better

1. The Earth-to-spacecraft distance had increased from approximately 5 AU at Jupiter to approximately 10 AU at Saturn.

2. Typically, the power of the X-band radio signal received by the DSN 64-meter antennas was 10^{-14} watt (-170dBm), unimaginably small.

3. The incredibly complex task of reprogramming the computers on the Voyager spacecraft, which were then traveling between Saturn and Uranus, was an engineering triumph in its own right. It was accomplished by transmitting long sequences of updated software commands from the DSN transmitters to the spacecraft over a period of many weeks, verifying that they had been received correctly by the spacecraft, and finally instructing the spacecraft to reprogram its computers with the new software.

4. Implementation of the Parkes-Canberra Telemetry Array was also a technological marvel, since the Parkes and Canberra antennas were more than two hundred miles apart. For a detailed account of how this was done, see Mudgway (2001).

5. Ka-band refers to the radio frequency band around 32 GHz having a wavelength of about 0.94 cm.

6. "Radio frequency (RF) figure of merit" is the term used by microwave antenna engineers to completely describe the radio frequency performance, or degree of "goodness," of a microwave antenna. The two key parameters on which it is based are the gain and the noise temperature of the antenna. Antenna engineers, like those at JPL, constantly strive to design antennas with high gain and low noise. Such antennas are blessed with a high RF figure of merit.

7. Antenna engineers call this function "illuminating" the dish with microwave energy. The more evenly the subreflector "illuminates" the dish, the more efficient the an-

tenna becomes as a collector of very weak microwave energy from a distant spacecraft or radio star.

8. At 280,000 pounds, each counterweight was equivalent to the takeoff weight of a fully loaded Boeing 767 airliner.

9. The reflector surface of the original 64-meter antenna contained 552 panels, but they were not of the same sizes as those required for the 70-meter antenna.

10. Because the surface contour of a precision microwave antenna changes slightly as the antenna moves in elevation, antenna designers define an elevation angle to which all contour measurements can be referred. This is termed the rigging angle. DSN antenna engineers use a rigging angle of 45 degrees for this purpose.

11. See McGinness (1985). This is one of the four articles that covered the McGinness study. They all appeared in the same issue of the Progress Report.

12. The tricone that was fitted to the 64-meter antenna was not changed in the 70-meter upgrade project. The new feeds were therefore required to be compatible with the mounting locations on the existing tricone.

13. The figure 22 dBi refers to the RF gain of the horn as 22 dB with respect to an isotropic radiator.

14. The RMS value is implied wherever an error or accuracy figure is quoted in the discussion that follows.

15. These were the design values for the finished subreflector as mounted on the quadripod. The raw casting from which it was machined was much larger.

Chapter 9. Implementation

1. Silica gel is the white crystalline substance that is often found in tiny packets inserted in containers to keep their contents dry. It has a remarkable ability to absorb moisture from its surroundings.

2. "MOD E" is an engineering term used to quantify the elasticity, or stiffness, of materials such as concrete.

3. Later McClure was to discover that the California Department of Highways had first recognized the reactive aggregate problem in the 1930s but, because it affected only the elasticity of their structures—roads, bridges, culverts—and not their strength, little was done about it at that time.

4. The perimeter of the haunch, 250 feet in length, was divided into nine equal sections, and concrete was removed from three opposing sections at a time.

5. The results of laboratory tests taken at twenty-eight days and sixty days after curing showed MOD E values well in excess of the specified value of 5 million psi.

6. The design value of MOD E for the pedestal concrete was 5 million psi. The MOD E value for the original concrete had deteriorated to 2 million psi—way below specification.

7. The number 1,289 included seventeen spare panels, one of appropriate size and shape for each of the seventeen rows of the finished surface.

8. McClure commented, "Spain had rigorous limits on overtime (100 hours per year). Also, when work was scheduled for weekends or holidays, less than half of the workers showed up, and productivity for those that did work was very low."

9. Unique to JPL management style, Tiger Teams were established on an emergency basis to solve critical technical problems in complex deep space mission systems. Highly focused and limited in duration, Tiger Teams were fully empowered by top-level management to draw on whatever JPL resources were necessary to solve problems as quickly as possible. Spacecraft emergencies and DSN antenna outages were typical examples.

10. The use of an "interim" subreflector using composite materials was a cause for great concern at the time because of the significant risk of overheating, or even fire, when the plastic and aluminum mesh materials were subjected to the intense microwave energy radiated by the 400-kilowatt uplink transmitter. That the final product withstood these severe conditions without evidence of stress attests to the particular attention that was given to the details of its design and construction by JPL's expert consultant, John Warren.

11. In addition to meeting its design goal for improvement in gain relative to the 64-meter antennas, the 70-meter version of the Big Dish also demonstrated a new level of performance for the "area efficiency" of large antennas—76 percent at S-band and 70 percent at X-band.

Chapter 10. Closing Scenes

1. Coordinated Universal Time (UTC) is more commonly known for general time-keeping purposes as Greenwich Mean Time (GMT).

2. JPL announced the spacecraft's demise in a news release on September 21, 2003: "The *Galileo* spacecraft's 14-year odyssey came to an end on Sunday September 21 when the spacecraft passed into Jupiter's shadow and then disintegrated in the planet's dense atmosphere at 11:57 a.m. Pacific Daylight Time. 'We haven't lost a spacecraft; we've gained a stepping-stone into the future of space exploration,' said Dr. Torrance Johnson, *Galileo* project scientist."

3. Mars *Pathfinder* could return telemetry data to the Big Dishes at a maximum data rate of 11,060 bits per second under favorable RF communication link conditions.

4. Data volume returned is the product of data rate and antenna viewing-time. The higher the data rate, the less time a DSN antenna requires to capture a given volume of data from a spacecraft. A single DSN antenna, therefore, can service more spacecraft per unit of time at higher data rates than at lower data rates for the same data volume returned.

Bibliography

Abbreviations

DSN	Deep Space Network
JPL	Jet Propulsion Laboratory, California Institute of Technology, Pasadena, Calif.
NASA	National Aeronautics and Space Administration
OTDA	Office of Tracking and Data Acquisition
PR	Progress Report
TDA	Tracking and Data Acquisition
TM	Technical Memorandum
TMO	Telecommunications and Mission Operations
TR	Technical Report

Bartos, K. P., et al. "Implementation of the 64-Meter-Diameter Antennas at the Deep Space Stations in Australia and Spain." TM 33-692, JPL, January 15, 1975.

Bathker, Dan A. "Radio Frequency Performance of a 210-ft Ground Antenna: X-Band." TR 32-1417, JPL, December 15, 1969.

———. Interview with J. Alonso. Oral History of the DSN Program, Archives and Records Facility, JPL, June 1992.

Bluth, John F. Interview with the author. JPL, February 2000.

Briskman, Robert D. "Trip Report to Green Bank West Virginia on November 4, 1959." Memorandum to Space Flight Operations files, December 29, 1959. NASA History Office Archives, series "NASA Administration and Organization," subject "OTDA."

———. "Meeting on the Advanced Antenna at JPL, 9 November, 1960." Memorandum to Assistant Flight Director, Space Flight Operations, November 22, 1960. NASA History Office Archives, series "NASA Administration and Organization," subject "OTDA."

Buckley, Edmond C. "Use of Large Radio Telescopes in West Virginia by NASA." Memorandum to Director, Space Flight Operations, November 18, 1959. NASA History Office Archives, series "NASA Administration and Organization," subject "OTDA."

———. "Implementation of Large Antenna." Letter to Dr. Eberhardt Rechtin, August 2, 1960. NASA History Office Archives, series "NASA Administration and Organization," subject "JPL."

———. "Status of Goldstone 210-foot Antenna." Letter to Deputy Administrator, NASA, March 24, 1967. NASA History Office Archives, series "NASA Administration and Organization," subject "JPL."

Bugg, Sarah J. Correspondence with the author. Canberra Deep Space Communications Complex, Tidbinbilla, ACT, Australia, August 2001.

Burrows, William E. *This New Ocean: The Story of the First Space Age.* New York: Random House, 1998.

Butrica, Andrew J. *To See the Unseen: A History of Planetary Radar Astronomy.* NASA SP-4218. Washington, D.C.: NASA, 1996.

———, ed. *Beyond the Ionosphere: Fifty Years of Satellite Communication.* NASA SP-4217. Washington, D.C.: NASA, 1997.

Casperson, Ronald D. Interview with the author. Pasadena, Calif., August 2000.

Chamarro, Agustín. Personal communication with the author. Madrid, Spain, August 2001.

Corliss, William R. "A History of the Deep Space Network." NASA CR-151915. May 1, 1976.

Drake, Frank. "Our Weakness in Space." *Saturday Review,* January 2, 1965.

Edwards, C. D., Jr., C. T. Stelzried, L. J. Deutsch, and L. Swanson. "NASA's Deep-Space Telecommunications Road Map." TMO PR 42-136, JPL, February 15, 1999. http://tmo.jpl.nasa.gov/tmo/progress_report/42-136/136B.pdf.

Ghigo, Frank. "NRAO–Green Bank, West Virginia." www.gb.nrao.edu/fgdocs/oldscopes/html (accessed June 2004).

Goldstein, Richard, et al. "The Superior Conjunction of *Mariner IV.*" TR 32-1092, JPL, April 1, 1967.

Gore, Rick. "Neptune: *Voyager*'s Last Picture Show." *National Geographic,* August 1990.

Hall, J. R. "Radio Frequency Figure of Merit Enhancement Study for 64-m Antennas." DSN document 890-47, JPL, June 1975.

Hall, R. Cargill. *Lunar Impact: A History of Project Ranger.* Washington, D.C.: NASA, 1977. http://history.nasa.gov/SP-4210/pages/TOC.htm.

Hall, Robert D. Interview with the author. San Francisco, April 2000.

Hartley, R. B. "*Apollo 13* Mission Support." DSN Space Programs Summary 37-64, vol. 2, JPL, August 1970.

Ikard, W. L. "Conference on Large Aperture Antennas for Deep Space Communications." Memorandum for Space Flight Operations files, November 23, 1959. NASA History Office Archives, series "NASA Administration and Organization," subject "OTDA."

Jet Propulsion Laboratory (JPL). "Advanced Antenna System: Structural and Drive Components." DSN Space Programs Summary 37-31, vol. 3, JPL, January 1965a.

———. "DSIF: Robledo." TM 33-258, JPL, 1965b.

———. "DSIF: Tidbinbilla." TM 33-207, JPL, 1965c.

———. "The NASA/JPL 64-Meter-Diameter Antenna at Goldstone: Project Report." TDA TM 33-671, JPL, July 1974.

———. "Pathfinder Team Paints an Earth-like Picture of Early Mars." News release 97-89, JPL Public Information Office, October 9, 1997.

Keeling, Patricia Jernigan, ed. *Once Upon a Desert.* Barstow, Calif.: Mojave River Valley Museum Association, 1976.

Koppes, Clayton R. *JPL and the American Space Program: A History of the Jet Propulsion Laboratory.* New Haven: Yale University Press, 1982.

Layland, J. W., and D. W. Brown. "Planning for VLA/DSN Arrayed Support for the Voyager at Neptune." TDA PR 42-82, JPL, August 15, 1985.

Layland, J. W., P. J. Napier, and A. R. Thompson. "History of the NASA Stations in Australia." TDA PR 42-82, JPL, August 15, 1985.

Leslie, Robert A. Personal correspondence with the author. Canberra, ACT, Australia, October 1980.

McClure, Donald H. "Repair of the DSS 14 Pedestal Concrete." TDA PR 42-81, JPL, May 15, 1985.

———. Interview with the author. Port Townsend, Wash., August 1999a.

———. "The 64-m Antenna Rehabilitation and Performance Upgrade Project Report." Correspondence with the author, August 1999b.

McClure, D. H., and F. D. McLaughlin. "64-Meter to 70-Meter Antenna Extension." TDA PR 42-79, JPL, December 15, 1984.

———. "64-Meter Antenna Rehabilitation and Performance Upgrade Project Report." Correspondence with the author, October 2001.

McCurdy, Howard E. *Faster, Better, Cheaper: Low-Cost Innovation in the U.S. Space Program.* Baltimore: Johns Hopkins University Press, 2001.

McGinness, Houston D. "Elevation Bearing Maximum Load, 70-m Antenna." TDA PR 42-80 for October–December 1984, JPL, February 15, 1985.

———. Interview with J. Alonso. Oral History of the DSN Program, Archives and Records Facility, JPL, September 1992.

McLaughlin, Fred D. Interview with the author. Sonoma, Calif., September 1999.

Massey, Ed B. "Voyager: The Interstellar Mission." JPL, 2003. http://voyager.jpl.nasa.gov/mission/interstellar.html.

Merrick, Beth. Correspondence with the author. April 14, 2000.

Merrick, W. D., R. Stevens, and E. Rechtin. "Project Description, Advanced Antenna System for the Deep Space Instrumentation Facility." Engineering Planning Document (EPD) 5, rev. 4, JPL, April 1963.

Merrick, William D. "Deep Space Network, Space Programs Summary." Vol. 3, no. 37-31:88, JPL, January 1965.

———. Interview with J. Alonso. Oral History of the DSN Program, Archives and Records Facility, JPL, May 1992.

Moon, Germaine L. Ramounachou. *Barstow Depots and Harvey Houses.* Barstow, Calif.: Mojave River Valley Museum Association, 1980.

Morrison, Neil C. "Our History, Then and Now." Fort Irwin, Calif.: National Training Center and 11th Armored Cavalry Regiment Museum, 2004. http://www.11thacr.org/museum/NTC tour.html.

Mudgway, Douglas J. *Uplink-Downlink: A History of the NASA Deep Space Network, 1957–1997.* NASA SP-2001-4227. Washington, D.C.: NASA, 2001.

Murray, Bruce. *Journey Into Space: The First Three Decades of Space Exploration.* New York: W. W. Norton, 1989.

National Aeronautics and Space Administration (NASA). "Feasibility Study for Deep Space Antenna." News release 61-191, August 25, 1961.

———. "NASA Selects Rohr Corporation for New 210-Foot Antenna." News release 63-13, January 25, 1963.

———. "Goldstone Dedication, April 29, 1966." News release 66-88, April 26, 1966.

———. "210-ft Antennas, Spain and Australia." NASA NEWS release 68-213, December 12, 1968.

———. "210-ft Antennas Contract." NASA NEWS release 69-98, June 27, 1969.

National Radio Astronomy Observatory (NRAO). "About NRAO." www.nrao.edu/about.

Newcott, William R. "In the Court of King Jupiter." *National Geographic,* September 1999.

O'Neill, W. J., N. E. Ausman Jr., T. V. Johnson, M. R. Landano, and J. C. Mar. "Performing the Galileo Jupiter Mission with the Low-Gain Antenna (LGA) and an Enroute Report." IAF-93.Q.5.4.411. Paper presented at the 44th Congress of the International Astronautical Federation, Graz, Austria, October 16–22, 1993.

Pay, Rex. "Advanced Deep-Space Dish Readied." *Missiles and Rockets,* May 2, 1966.

Perminov, V. G. *The Difficult Road to Mars: A Brief History of Mars Exploration in the Soviet Union.* Monographs in Aerospace History 15. Washington, D.C.: NASA, 1999.

Pickering, William H. Letter to Brig. Gen. W. A. Jensen, CO, Camp Irwin, Calif., March 27, 1958. Document 3-1260, History Collection, Archives and Records Facility, JPL.

Potter, P. D. "The Application of Cassegrainian Principles to Ground Antennas for Space Communications." *IRE Transactions on Space Electronics and Telemetry* SET-8, no. 2 (June 1962).

Rechtin, E. "Long-Range Planning for the Deep Space Network." AIAA Paper 67-975. Presented to American Institute of Aeronautics and Astronautics, Anaheim, Calif., October 23, 1967.

Rechtin, Eberhardt, Bruce Rule, and R. Stevens. "Large Ground Antennas." TR 32-213, JPL, March 1962.

Renzetti, N. A., et al. "The Deep Space Network—A Radio Communications Instrument for Deep Space Exploration." Publication 82-104, JPL, July 15, 1983.

Siddiqi, Asif A. *Deep Space Chronicle: A Chronology of Deep Space and Planetary Probes, 1958–2000.* Monographs in Aerospace History 24. Washington, D.C.: NASA, 2002.

Smith, Bradford A. "Voyage of the Century." *National Geographic,* August 1990.

Stelzried, C. T., G. S. Levy, and M. S. Katow. "Multi-feed Cone Cassegrain Antenna." U.S. patent 3,534,375, October 1970.

Stelzried, Charles T. E-mail communication with the author. September 30, 2003.

Stevens, Robertson. "Summary of the work done on the Advanced Antenna System, 11/15 to 12/15, 1961." Letter to Gerald M. Truszynski, December 21, 1961. History Collection, folder 7-66, JPL.

———. "Applications of Telemetry Arraying in the DSN." TDA PR 42-72 for October–December 1982, JPL, February 1983.

———. "Implementation of Large Antennas for Deep Space Mission Support." TDA PR 42-76, JPL, February 1984.

Swaim, Dave. "Goldstone's New Space Ear Dedication Attended by 400." *Pasadena Star-News,* April 30, 1966.

Tardani, Phillip A. "Madrid Site Selection Report." TM 33-149, JPL, July 1963.

United Press International (UPI). "Plan Giant Antenna to Receive Moon TV." *Los Angeles Herald-Examiner,* March 8, 1962a.

———. "Minuteman Lands on Target 3,000 Miles Away." *Los Angeles Times,* March 9, 1962b.

"Voyage to the Morning Star." *Time,* March 8, 1963.

Wells, I. Dale. Personal correspondence and interview with the author, Nippomo, Calif., July 1999.

Wood, Richard D. "Report on the Mechanical Maintenance of the 70-Meter Antennas." Document D-10442, JPL, January 1993.

Index

Page numbers in *italics* refer to photographs and illustrations.

A mathematics and physics graduate of the University of New Zealand (1945), Douglas Mudgway came to the United States in 1962 to work at NASA's Jet Propulsion Laboratory in Pasadena, California, following a fifteen-year career in the field of guided missile research and testing at Woomera, Australia. At JPL he was involved in the development and operation of the Deep Space Network from its infancy in the early 1960s to its maturity in the early 1990s.

In 1977 he was awarded the NASA Exceptional Service Medal for his contribution to the first Viking Lander mission to Mars, and he received the NASA Exceptional Achievement Medal in 1991 for his contribution to the Galileo mission to Jupiter.

Following retirement from active JPL service in 1991, he acted as an independent consultant on deep space planetary communications for NASA/JPL and has written extensively on that subject. His previous book, *Uplink-Downlink: A History of the NASA Deep Space Network, 1957–1997,* was published by NASA in 2001.

Titles of related interest from the University Press of Florida

Stages to Saturn: A Technological History of the Apollo/Saturn Launch Vehicles
Roger E. Bilstein

The Soviet Space Race with Apollo
Asif A. Siddiqi

Sputnik and the Soviet Space Challenge
Asif A. Siddiqi

Gateway to the Moon: Building the Kennedy Space Center Launch Complex
Charles D. Benson and William B. Faherty

Moon Launch! A History of the Saturn-Apollo Launch Operations
Charles D. Benson and William B. Faherty

Florida's Space Coast: The Impact of NASA on the Sunshine State
William Barnaby Faherty, S.J.

"Before This Decade Is Out...": Personal Reflections on the Apollo Program
Glen E. Swanson

For more information on these and other books, visit our website at www.upf.com.